MARIANNE TAYLOR &
DANIELE OCCHIATO

BIRDS OF ITALY

A PHOTOGRAPHIC GUIDE

SECOND EDITION

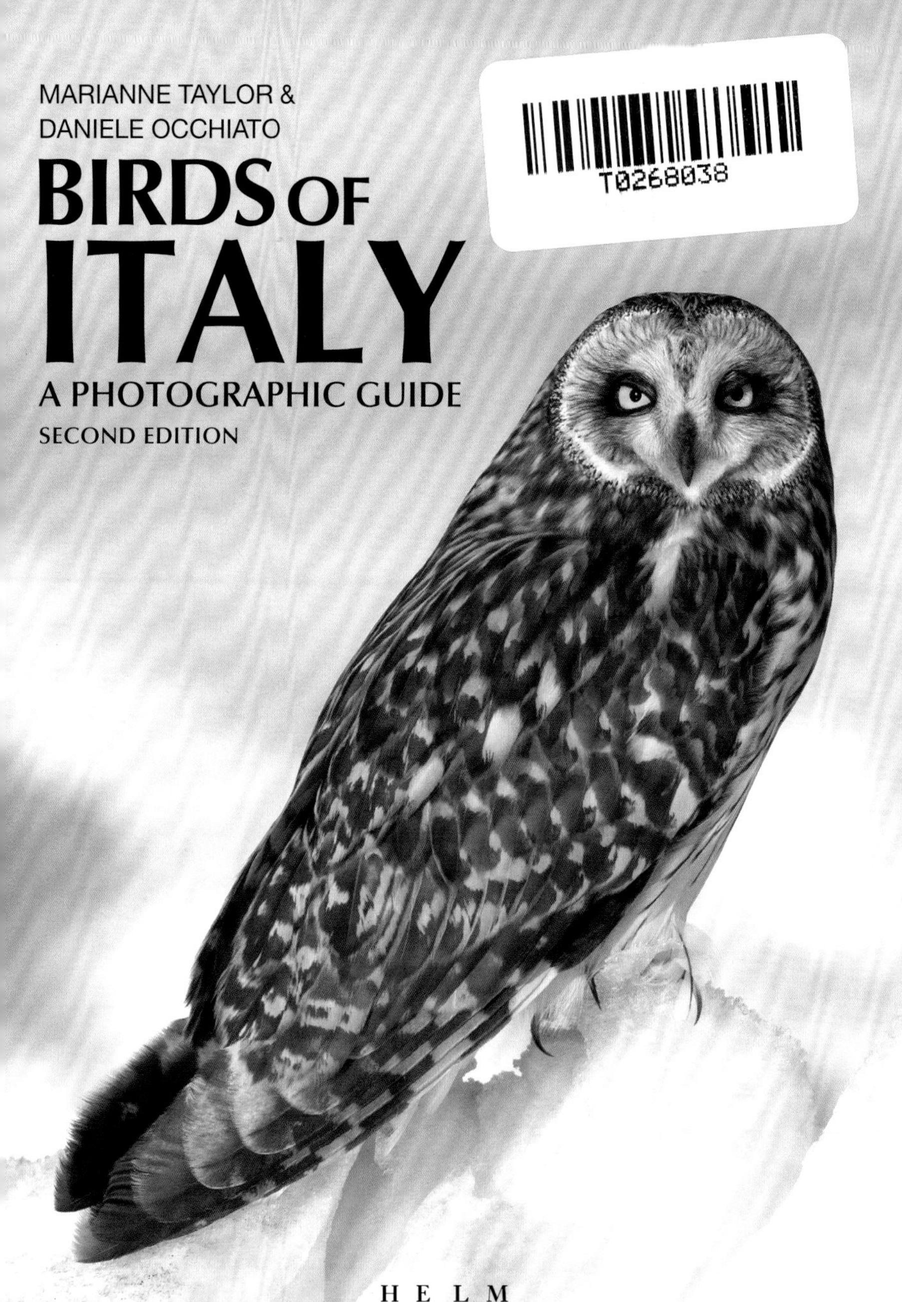

H E L M

LONDON • OXFORD • NEW YORK • NEW DELHI • SYDNEY

HELM
Bloomsbury Publishing Plc
50 Bedford Square, London, WC1B 3DP, UK
29 Earlsfort Terrace, Dublin 2, Ireland

This edition published 2024

A catalogue record for this book is available from the British Library.

Library of Congress Cataloguing-in-Publication data has been applied for.

ISBN: PB: 978-1-3994-1064-9; ePDF: 978-1-3994-1066-3;
ePub: 978-1-3994-1063-2

2 4 6 8 10 9 7 5 3 1

Design by Rod Teasdale
Map by Julie Dando

Printed and bound in Türkiye by Elma Basim

CONTENTS

Introduction	4
Habitats	5
Birdwatching in Italy	6
Map of the region	7
Best birdwatching sites in Italy	8
How to use this book	19
Species accounts	20
Glossary	220
Further reading	220
Photo credits	221
Acknowledgements	221
Index	222

INTRODUCTION

Italy marks the transition from south-western to south-eastern Europe. The country comprises the mainland with its elegant high-heeled boot outline, which stretches south-east into the Mediterranean, along with the major islands of Sicily and Sardinia (plus a few groups of smaller islands). With its generally warm climate, varied and spectacular scenery, famous cityscapes and unique archaeological wonders, Italy is an extremely popular destination for tourists.

The history of Italy's wild places and wildlife has not been an entirely happy tale. Rapid industrialisation led to pollution, forest clearance, rapid development of formerly wild coastlines and the loss of vast tracts of habitats, while efforts to reduce the hunting of wild birds has met with strong resistance from a powerful gun lobby. However, much of Italy's wild beauty has survived intact, and conservation progress since the 1990s has been considerable. There are now 25 national parks in Italy and these, along with other protected areas, cover more than 11% of the total land area. One third of animal species native to Europe can be found in Italy, and about 540 bird species have been recorded here – well over half of all species recorded in Europe as a whole, although many of these are 'vagrants' rather than regular visitors. This book covers the most frequently encountered or distinctive birds of Italy – 330 species in all. For an exhaustive guide to all species ever observed in Italy, we recommend the *Collins Bird Guide* (2022) – see Further reading on page 220.

Glossy Ibis

HABITATS

Italy's shape and location lend it considerable variation in climate from north to south. The land profile also varies a great deal – much of the mainland is hilly or mountainous but there are also some important lowland wetlands, and more than 9,000km of coastline. The two major islands, Sicily and Sardinia, have their own particular geographic character and wildlife.

Mountains Italy's northern edge is marked by the Alps, Europe's highest and most significant mountain range. A few cold-climate montane species, such as Ptarmigan, occur only here, while other mountain-dwelling birds can also be found in the Apennines further south, in the uplands of the south, and on Sicily and Sardinia.

Forest There are significant remnants of primeval deciduous forest in Calabria and Tuscany, while in the alpine regions are forests of fir and pine, home to Capercaillies, Nutcrackers and many more birds. The Apennine foothills support woodlands of Neapolitan maple and Italian alder. On Sicily, Sardinia and the southern mainland, woodland is dominated by Mediterranean species such as cork oak and Aleppo pine.

Scrub/maquis Where forest has been cleared, or the soil does not support its growth, a mosaic of bushes and tall herbaceous plants may grow. This is rich in insect life and accordingly supports a great range of bird species, including various species of shrikes and *Sylvia* and *Curruca* warblers. Maquis scrub, consisting mainly of thick evergreen shrubs like broom and myrtle, is common in the south and on Sicily and Sardinia.

Meadows and cultivated land Untouched grassland is scarce in Italy but alpine meadows offer great hunting grounds for birds and other wildlife, especially butterflies. Vineyards, olive groves, grazing land and other farmland can also be productive for open-country and scrubland birds.

Rivers and lakes Italy has few major rivers, the longest being the 650km- long Po, with its source in the western Alps. Upland rivers and streams support their own specialist birds, such as Dippers, while lowland rivers attract a variety of wetland species, and European Bee-eaters, Kingfishers and Sand Martins nest in riverbanks. The high, steep-banked glacial lakes support rather few birds, but well-vegetated lowland lakes attract many species.

Marshland The most important area of marshland in Italy is the Po delta in the north-east, but smaller marshes, lagoons and wetlands can be found elsewhere, for example in Circeo National Park on the Lazio coast. They support many breeding birds and offer rest and respite for migrant waders and wildfowl.

Coasts, islands and offshore Much of the Italian coastline has been developed for tourism, but there remain some unspoilt stretches of rocky shoreline and extensive sand dunes. Tiny rocky islets provide breeding sites for special birdlife including Eleonora's Falcon and Scopoli's Shearwater, while some seabirds winter offshore around the Mediterranean Sea and may be observed from headlands.

Urban environments Town parks and gardens often support more common woodland birds, and colonies of exotic feral birds like Rose-ringed Parakeets. Archaeological sites and older settlements can be wonderful places to seek out birdlife – for example, the beautiful walled city of Matera in Basilicata, with its famous Lesser Kestrel colony.

BIRDWATCHING IN ITALY

Italy has many national parks, nature reserves and other protected areas, and also a number of long-distance hiking trails that traverse some of the most spectacularly scenic parts of the country. In the countryside in general, you'll find many paths and tracks with open access, and locals will usually be happy to offer advice if needed. In short, exploring wild places in Italy is very safe and straightforward, although venturing onto the more challenging alpine and other mountain trails will require a good level of fitness, as well as careful preparation and common sense. Take the usual precautions when out for a walk – carry water, sun protection and a charged mobile phone.

Walking maps from Tabacco Casa Editrice cover many areas, and there are many English-language regional walking guides available, with recommended routes mapped and described in detail. Some national parks also sell their own maps and guides. Nature reserves and national parks will usually have waymarked trails for you to follow, and some have hides for closer views of key spots. You will need binoculars, and a telescope will be useful, particularly at the coast and in the mountains.

Hunting is widespread in Italy, and the hunting season is long, beginning on the first Sunday in September and continuing to the end of February or March depending on the quarry species (until the end of January for birds). It is prohibited in some but not all protected wild places – be wary of leaving main hiking routes during the open season, and some sites are best avoided altogether at this time as they will simply not offer productive birding. Illegal killing of protected species, particularly raptors, is still a frequent problem. If you see anything that raises concerns, contact LIPU (BirdLife in Italy, www.lipu.it) for advice. As in other countries with a long hunting tradition, you may find that birds are very wary of humans and close approach may be difficult, although this will vary from region to region (and some species seem naturally fearless anyway).

Many birding tour companies offer trips to Italy, particularly to regions such as the Po delta, Alps and Sicily. This is the easiest way for a newcomer to Italian birding to connect with the more unusual species, while more general-interest tours will usually also offer some birding opportunities. Independent travellers may wish to contact a local birder for a more informal arrangement – try the Birding Pals website: www.birdingpal.org/Italy.htm.

MAP OF THE REGION

BEST BIRDWATCHING SITES IN ITALY

Italy is full of places to explore, and this section suggests some of the best and most accessible birdwatching sites in the country. There are, of course, many more for the adventurous traveller to discover. Explore information on protected sites in Italy here: www.parks.it/Eindex.php. The website www.birdingplaces.eu offers detailed information on the sites listed below, and many more. The following sites are listed by region.

ABRUZZO

Gran Sasso National Park – Monte Aquila

'Eagle Mountain' has a fairly tough 10km circular trail, which may reward you with views of both species of choughs, Alpine Accentor, White-winged Snowfinch, Common Rock Thrush, Water and Tawny Pipits and Peregrine Falcon, as well as the eponymous Golden Eagle. Other, shorter trails are also available in the park.

La Camosciara

Set within the Parco Nazionale di Abruzzo Lazio Molise, this area of upland and woodland is home to White-backed Woodpecker, Dipper, Red-billed Chough, Raven and Golden Eagle.

AOSTA VALLEY

Champoluc

This pleasant walk south of Champoluc takes you through open meadowland, with forest above and the Evancon river below. Spring is the best time to visit, when migrants such as Pied Flycatcher, Wryneck and various warblers join the breeding species, among them Willow and Crested Tits and Firecrest.

Valsavaranche

A wild mountain valley with an array of rugged paths to climb. Here you could see Bearded Vulture, Rock Partridge, Ptarmigan, Golden Eagle, Wallcreeper, Nutcracker and other typical montane northern species.

Valsavaranche, Aosta Valley

Valle Argentera

An alpine valley which you can explore by road or on foot. Look out for Bearded and Griffon Vultures, Golden Eagle, Eagle Owl and other upland birds.

APULIA

Gravina de Laterza

This LIPU reserve has ravines and caves, and offers great walking and cycling paths. Here you can see Short-toed Eagle, Eastern

Black-eared Wheatear, Egyptian Vulture, Black Stork and Lanner Falcon.

Oasi Laguna del Re
This wetland site adjoins other reserves and sits on a migratory flightpath for many species. Look out for Greater Flamingo, Spoonbill, various herons, and godwits and other waders.

Parco Nazionale Alta Murgia
With its wild open plains, this national park offers a kind of scenery that is scarce in Italy. A key area for Lesser Kestrel, also steppe birds such as Eastern Black-eared Wheatear and Calandra Lark.

Ponte Ciolo
This rocky canyon has a range of bird species typical of Mediterranean maquis vegetation, as well as a rocky shoreline with sea caves. You could find gulls including Slender-billed and Audouin's, and also White and Black Storks and Booted Eagle, with warblers including Sardinian in the scrubby areas.

Riserva Statale Torre Guaceto
A large and important area of Mediterranean maquis scrub; habitat for shrikes, warblers and other scrubland species.

BASILICATA

Parco della Murgia Materana
This dramatic limestone landscape is a great place to watch raptors, particularly Lesser Kestrel (note that this species also breeds in the nearby city of Matera, within its spectacular World Heritage-designated Sassi district). Also present are Red Kite, Lanner Falcon, Blue Rock Thrush and Egyptian Vulture.

Parco Nazionale del Pollino
A large and wild mountainous area between the Basilicata and Calabria regions. It holds raptors including Red Kite and introduced Griffon Vulture, and landbirds including Black Woodpecker and Rock Partridge.

CALABRIA

Stretto di Messina – see Sicily

Lago di Tarsia
This reservoir has many attributes of a natural wetland, and accordingly supports birds such as Marsh Harrier, White Stork, various herons and grebes, and has breeding European Bee-eaters.

Lago dell'Angitola
A large lake surrounded by olive groves, woodlands and Mediterranean maquis. An important bird area for waterbirds at any season, with wintering waterfowl, and for breeding landbirds. Especially productive during spring migration.

Parco Nazionale della Sila
A montane park with very extensive coniferous and mixed woodlands, fields, pasturelands and large lakes. It holds woodpeckers, including Black and Middle Spotted, raptors, Rock Sparrow, Rock Bunting, Siskin, Collared Flycatcher, and many other passerines.

Marchesato
An area of open, dry hills with some rocky outcrops, olive groves, Mediterranean maquis and extensive crops around San Mauro Marchesato. A great area for Mediterranean birds, raptors and breeding Black Stork.

CAMPANIA

Fiume Tanagro Sala Consilina

Paths take you through open countryside along the Tanagro river. This area is noted for a breeding population of White Storks, but you can also find interesting farmland birds such as Crested Lark, Common Quail and Corn Bunting. Wetter areas are home to Marsh Harrier, Common Crane, and Great White and Cattle Egrets.

Foce del Fiume Sarno

North of the point where the Sarno river enters the sea, you can walk to a small headland and, if conditions are good, enjoy some great sea-watching with the chance of seeing Scopoli's and Yelkouan Shearwaters, and Gull-billed and Sandwich Terns. Waders may also pass through.

Gola del Diavolo del Fiume Mingardo

Along this stretch of the Mingardo river, the water flows fast through a steep-sided valley. Dippers may be seen here, along with Kingfishers and Common Sandpipers, with herons on sections with more gentle flow. Alpine and Common Swifts fly overhead.

Oasi WWF Persano

This reserve offers beautiful walks, interspersed with birdwatching hides from where you can view the wetlands. Look for breeding Ferruginous Duck, Squacco Heron and other waterbirds, and during migration seasons you could encounter waders such as Curlew Sandpiper.

EMILIA-ROMAGNA

Ex Risaia Di Bentivoglio

A nature reserve established on a former rice paddy, this wetland's shallow shores attract passage waders such as Black-tailed Godwit and Avocet, with ducks including Garganey and Gadwall on the open water. Several heron species breed, including Night Heron.

Parco dei Gessi Bolognesi e Calanchi dell'Abbadessa

This reserve comprises many different habitats including cliffs, forest, scrub and open countryside. Exploring the paths (including a 4km birdwatching trail) could bring you sightings of Moltoni's Warbler, Melodious Warbler, Red-backed Shrike, Montagu's Harrier, Tawny Pipit, European Bee-eater and Cirl Bunting.

Parco Naturale del Delta del Po

An extensive, spectacular area of marshland and open water, home to Gull-billed Tern, Ferruginous Duck and many more wetland birds.

Valli Mirandolesi

This series of valleys and wetlands is protected and offers refuge to interesting breeding birds such as Whiskered Tern (one of the best sites in Italy for this species), Great Bittern, Purple and Squacco Herons, Black-winged Stilt, Avocet and Kentish Plover. Breeding wildfowl include Shoveler and Garganey, and there is also a good variety of songbirds including Bluethroat, and Savi's and Great Reed Warblers.

FRIULI-VENEZIA GIULIA

Piana del Preval

This marshland valley, with wooded hills around it, is a Special Area of Conservation, readily accessible on foot.

The diverse habitats here provide the chance to see many species, including Long-eared and Eurasian Scops Owls, Osprey, Corncrake, Penduline Tit and many species of ducks and waders.

Riserva Naturale della Foce dell'Isonzo
Situated in the eastern part of the Po delta, this wetland nature reserve has superb viewing facilities and attracts large numbers of waders, wildfowl and herons.

Sorgenti della Santissima
Freshwater springs here give rise to the Livenza river. The area is popular with walkers and the trails are easy. The woodlands here may provide sightings of Black and Lesser Spotted Woodpeckers, European Honey-buzzard and, on the stream, Dipper.

LAZIO

Palude di Torre Flavia
A coastal marshland, good in the breeding season for Little Bittern, Moustached Warbler and other wetland birds. From the beach look out for seabirds including Scopoli's Shearwaters, and in winter visiting waders such as Kentish Plover. The meadows further inland may produce Yellow Wagtail and Zitting Cisticola.

Parco di Gianola
This protected area of coastal and woodland habitat is a great spot to visit during migration, to see an array of warblers and other songbirds as well as seabirds.

Parco Nazionale del Gran Sasso e Monti della Laga
In the heart of the Apennines, this is one of Italy's largest national parks, with exceptional botanical biodiversity, and birdlife including Golden Eagle, Wallcreeper and White-winged Snowfinch.

Parco Nazionale del Circeo, Lazio

Parco Nazionale del Circeo
Formed of a great crescent-shaped dune, ancient coastal forest and a series of lakes surrounded by marshland, this picturesque national park supports a great range of birdlife.

Monti della Tolfa
A rather wild and extensive area of low hills with Mediterranean woodlands and maquis, open rocky pastures and small outcrops. It is a key area for raptors and landbirds with breeding Lesser Kestrel, Short-toed Eagle, Red Kite, Black-headed Bunting, Calandra Lark, Spectacled warbler, Roller and European Bee-eater.

LIGURIA

Cima Ramà
Migrating raptors cross between high mountains at this point on the Mediterranean coast. In both migration seasons, you can see good numbers of Short-toed Eagles, along with European Honey-buzzards. Resident birds include Griffon Vulture and Blue Rock Thrush, and Wallcreepers may winter.

Monte Reixa

Part of the Parco Naturale Regionale del Beigua, this mountain has a relatively easy walking trail which takes you through forest and out onto rocky pasture. Look out for Crested Tit and Pied Flycatcher in the woodland, while the open areas are good for spotting raptors such as Short-toed and Booted Eagles, as well as Tawny Pipit and, in spring, the chance of migrating Dotterels.

Oasi del Nervia

The estuary of the Nervia river is a protected area, which you can view from surrounding paths and a bridge across the river. Breeding species present include Turtle Dove, Baillon's Crake and Bluethroat, with many wildfowl visiting at other times. Head to the coast to look for divers, mergansers and shearwaters offshore.

Punta Chiappa

A headland within the Portofino Natural Park. This site is excellent for seawatching, with seabirds often close to shore. They include shearwaters and gulls as well as Razorbill in winter. The sea cliffs are home to Blue Rock Thrush and Wallcreeper.

Punta Chiappa, Liguria

LOMBARDY

Brebbia fields

This area of farmland is a particular hotspot for passerine migration in spring and autumn, with several species stopping to feed, sometimes in large numbers. It is a regular site for scarcer species such as Red-throated Pipit and Short-toed Lark, as well as numerous Yellow Wagtails, Water Pipits and assorted finches and buntings.

Oasi La Volano

A rewilded haven in an agricultural desert, this wetland has a viewing platform across open water and marshland. An impressive sightings list includes Ruddy Shelduck and Red-crested Pochard, Penduline Tit, several heron species and waders such as Common Snipe and Black-winged Stilt.

Oasi Lipu di Cesano Maderno

A diverse LIPU reserve, including wetland, woodland and open moor. It is home to a range of waterfowl including Little Bittern, also woodland species including European Honey-buzzard.

Parco Cassinis

This is the best of Milan's parks for birdwatching, with its wooded areas and open spaces providing homes for the likes of Golden Oriole, Melodious Warbler, Nightingale and Hobby.

Parco Le Folaghe

Follow the 3km trail to explore this wetland habitat which, isolated as it is within an expanse of arable land, is something of a magnet for passing waterbirds. Visiting waders include Common and Wood Sandpipers, and you could also see several heron species including Squacco, as well as Turtle Dove, Common Cuckoo and Marsh Harrier.

Parco Nazionale dello Stelvio

An expanse of alpine mountains with more than 1,500km of paths; borders several other national parks. Montane and forest birds, including Golden Eagle, Eagle Owl and five woodpecker species.

Pian di Spagna e Lago di Mezzola

This nature reserve comprises Lake Mezzola, which adjoins Lake Como in the north-east. In winter the lake attracts many waterbirds including Slavonian Grebe, with visiting winter raptors such as Merlin, while the surrounding countryside holds summer visitors including Hoopoe, Golden Oriole, Pied Flycatcher and Western Orphean Warbler.

Torbiere del Sebino

A natural reserve in the Po valley, comprising an extensive area of marshland and reedbed. Species present include a large population of Savi's Warbler, also Great Reed Warbler and Penduline Tit.

Torbiere del Sebino, Lombardy

MARCHE

Monte Conero – Gradina del Poggio

The headland here is an exceptional site for watching raptor migration in spring, second only to the Stretto di Messina. Species that cross here in suitable conditions include European Honey-buzzard, Short-toed Eagle, Red-footed Falcon, and also both Black and White Storks and Common Cranes.

Monti delle Cesane

This is an area of mountainous pine forest, good for woodland birds including owls and woodpeckers, Goshawk and European Honey-buzzard. More open areas could provide sightings of Peregrine Falcon and Short-toed Eagle.

Monte Nerone

This 1,525m mountain offers good opportunities to watch birds of prey. You could find Golden Eagle, Goshawk and Griffon Vulture, among others, with Montagu's Harrier, Hobby and Little Owl present lower down.

Riserva Naturale Regionale Sentina

Rivers and lakes, marshes, dunes and meadows support numerous birds in a beautiful landscape. Look out for Zitting Cisticola, Golden Oriole, Serin, European Bee-eater and Turtle Dove.

PIEDMONT

Bossolasco

Around this village are upland meadows and woods, where you can walk several trails and encounter an interesting range of species. More notable birds present include Cirl Bunting, Crested Tit, Hawfinch, Common Crane, European Bee-eater, Crag Martin and Short-toed Eagle.

Bric Lobarera

A high-altitude watchpoint for migrating raptors, especially productive in autumn.

The most abundant species you will see is European Honey-buzzard, with Short-toed Eagle, Goshawk, Hobby and Black Kite also passing through – Black Storks are also regularly seen.

Maddalena pass
This open, high mountain pass provides opportunities to see a range of alpine species, including Bearded and Griffon Vultures, Alpine Chough and Golden Eagle.

Oasi La Madonnina
A wetland managed for wildlife and adjoining the Stura river, this site has breeding Eurasian Teal and a few other wildfowl species, but draws in many more during migration periods, including Pintail. Tern rafts provide nest sites for a population of Common Terns, and egrets and herons also breed. Raptors use the river as a landmark during migration, and good numbers of European Honey-buzzards may be seen, alongside Red-footed Falcons and Black Kites.

Oasi Naturalistica di Isola Sant'Antonio
This scenic restored wetland attracts sizeable populations of wildfowl on migration and in winter, including Red-crested Pochard, Ferruginous Duck, Pintail, Shoveler, Wigeon and Tufted Duck. Breeding waterbirds include Black-necked Grebe and Great Reed Warbler. The site is not normally open to the public but can be visited by reservation, and during special events.

Parco Nazionale Gran Paradiso
This vast national park holds the full suite of alpine habitat types and associated birdlife.

Riserva Naturale di Crava-Morozzo
A small nature reserve managed by LIPU, with screens and hides for closer views. Breeding species include Golden Oriole, Little Bittern, Night Heron and Black-winged Stilt.

Roggia Fonna
An extensive and varied wetland, especially good for waders and wildfowl during spring and autumn migration. Look out for herons, including Squacco, Hen Harrier and Black-winged Stilt, among many others. Passage waders include Spotted Redshank and Wood Sandpiper, while Bluethroat and Grasshopper Warbler breed.

SARDINIA

Foce del Cedrino
This narrow strip of shore extends across the mouth of the Cedrino river, creating a sheltered lagoon. This provides feeding grounds for Greater Flamingos, waders (especially plovers) and gulls, with Sardinian Warblers present in surrounding scrubland.

Monte Minerva
This plateau is the site of a highly successful reintroduction programme for Griffon Vultures. You can enjoy good views of the vultures at feeding stations, and other upland birds including Golden Eagles are also present.

Parco Naturale Regionale Molentargius
Extensive lagoons and marshland, with breeding Greater Flamingos and many other wetland birds.

Stagno di Porto Botte
A coastal lagoon with surrounding salt steppes, this area is easily explored and

should provide you with views of breeding Greater Flamingos and Black-winged Stilts. Little and Common Terns are also present in summer, and various wildfowl visit, along with Slender-billed Gulls.

Stagno Morto

Despite its discouraging name (meaning 'dead pond'), this site is a haven for breeding and migrating waders, with the likes of Kentish Plover, Black-winged Stilt and Greenshank visiting the reed-fringed coastal lagoons. Ospreys may winter in the area.

SICILY

Diga Rubino

This artificial reservoir in north-western Sicily is well-placed to attract passing migrant waterfowl and raptors, with interesting landbirds in the surrounding habitats. Look out for Woodchat Shrike, Common Cuckoo and Pallid Swift, with Ferruginous Duck among the various wildfowl on the reservoir.

Isola di Ustica

This 9km^2 volcanic island lies some 70km north of Sicily itself, and you can visit for the day or stay over. It is a great watchpoint for raptor migration in spring, with European Honey-buzzard topping a list that also includes Black Kite and Marsh Harrier. Storks also cross here, as do large numbers of European Bee-eaters.

Penisola Magnisi

An interesting environment of open rocky terrain with sparse vegetation, this peninsula is particularly worth a visit during spring migration. Migrant and resident passerines to be found here include Short-toed, Calandra and Crested Larks, Subalpine Warbler and Eastern Black-eared Wheatear. There are several good spots for seawatching – look out for passing Scopoli's Shearwaters.

Stretto di Messina

Italy forms a flyway for large, soaring migratory birds that prefer to avoid sea crossings, and many cross from the mainland to Sicily via this narrow strait, offering exciting spring and autumn birdwatching.

Riserva Naturale Integrale Lago Preola e Gorghi Tondi

One of several fine wetland sites on this stretch of coast. Ferruginous Duck, Purple Heron and Purple Swamphen are all present, and it is also a good site to observe raptors, including Booted Eagle.

Riserva Naturale Orientata Zingaro

A diverse and beautiful stretch of coastline, good for watching seabirds and finding scrubland species.

TRENTINO-SOUTH TYROL

Altopiano di Musiera

A peaceful expanse of pine and beech forest with areas of open meadowland, home to Capercaillie and Black Grouse, also Rock Partridge and Tengmalm's Owl.

Biotopo di Terlago

The easy walking trail around the southern of the two Terlago lakes offers opportunities to observe woodland and water-birds. Species include Night Heron, Short-toed Eagle, Crag Martin and Common Cuckoo.

Confluenza Adige-Noce

Where the Adige river meets the smaller Noce in a deep and lush valley, wetland birds gather on migration, while Eagle Owls and Golden Eagles breed on the cliffs above. Waders can be abundant, especially in autumn, and may include Greenshank, Wood Sandpiper and Green Sandpiper, while there is a sizeable gull roost in the area in winter. The wooded areas are home to Western Bonelli's and Wood Warblers in summer.

Fanes-Sennes-Prags

This breathtakingly beautiful little lake is a protected site within the Dolomites UNESCO World Heritage Site. Dippers may be seen here, although the main birding interest lies in the surrounding woods, with species including Black Woodpecker and Crested Tit present. Keep an eye on the skies, too, for Golden Eagle, Raven and Red Kite.

Lago de Tenno

This spectacular, intensely blue lake lies close to Lake Garda and is surrounded by forest. There is a circular path around the shore, and various trails through the woodlands. Tengmalm's Owl is present, though difficult to see, and you may spot Golden Eagles soaring overhead. Look out for upland waterbirds such as Common Sandpiper.

Lago de Tenno, Trentino-South Tyrol

Monte Tauro

A somewhat strenuous all-day hike in summer will take you around this scenic upland region and provide opportunities to see typical birds of the region, including Ptarmigan, Rock Partridge and perhaps Golden Eagle and Capercaillie.

Val de la Mare

A scenic valley surrounded by high mountains, this site packs a lot of habitat interest into a relatively small area. In the hillside woodlands are Tengmalm's and Pygmy Owls, while open alpine meadows are full of flowers and insects, and accordingly attract many passerines in springtime. At higher altitudes are Alpine Choughs, Rock Partridges and other montane species.

TUSCANY

Bocca d'Ombrone

A stretch of coastline, where the Ombrone river reaches the sea. Following the 4km trail takes you through shoreline and coastal habitats and is especially productive during migration seasons. Look out for shorebirds including Whimbrel and Spotted Redshank, and at sea you could spot Scopoli's and Yelkouan Shearwaters. In summer there is breeding Roller and Great Spotted Cuckoo while in winter you can easily find flock of Cranes and Greylag Goose, Curlew, Golden Plovers and Stone Curlew.

Isola di Capraia

A remote and beautiful island, part of the protected Arcipelago Toscano, with nesting Scopoli's Shearwater and Marmora's Warbler. Especially productive in April when almost every migrant could turn up.

Stagni della Piana Fiorentina
Located near Florence airport (Peretola), it is a plain with some small lakes, fields, small patches of trees and scrubland between the cities of Florence and Prato. The whole area is very good for waterbirds and landbirds, especially during migration. There are breeding Black-winged Stilt, various herons, Pygmy Cormorant and European Bee-eater.

Parco di Migliarino San Rossore, Massaciuccoli
This Tuscan natural park is a rare area of unspoilt coastline, comprising beach, lake, scrubland and forest, and is a key migration stopover as well as hosting numerous wintering and breeding birds. LIPU-protected Lake Massaciuccoli is a wetland reserve incorporating the lake and its surrounding extensive marshes and reedbeds; this site can be explored by boat as well as on foot. You could see Little Bittern, Black Tern, wetland warblers, Penduline Tit and Osprey.

Padule di Fucecchio
Italy's largest inland wetland. This nature reserve with a few hides holds a very large mixed heronry (up to 1,300 pairs of Night Heron, Little Egret, Squacco Heron, Cattle Egret, Grey Heron, Great White Egret, Purple Heron, Glossy Ibis, Sacred Ibis and Pygmy Cormorant). It is a vital area for wintering wildfowl and other waterbirds.

Riserva Naturale Bosco di Tanali
A very small lake surrounded by reedbeds and woodland. Easy accessible on foot. Here you will find passerines typical of wet woodland and marsh, including Cetti's and Reed Warblers, Nightingale, and Penduline Tit: the invasive Red-billed Leiothrix is particularly common here.

Orbetello Lagoon
An extensive coastal wetland with Mediterranean woodland and maquis, which includes the WWF Orbetello Nature reserve. There are breeding Greater Flamingo, herons, terns, waders, Roller, European Bee-eater, Great Spotted Cuckoo and many Mediterranean passerines. A key wintering area.

Padule della Diaccia Botrona
A very large coastal wetland surrounded by Mediterranean woods and maquis, and one of the most important humid areas of Italy. In winter there are thousands of waterbirds (Greater Flamingos, Spoonbills, Cranes, waterfowls, waders, and usually one or two Great Spotted Eagles). In summer, Roller, European Bee-eater and other Mediterranean passerines breed here.

Apuan Alps
High mountains with Alpine Accentor, both choughs species, Wallcreeper, Common Rock Thrush, raptors and other alpine passerines.

VENETO

Barruco
A road traverses this rich section of the Po delta, giving views across lagoons and marshy meadowland. A wide range of wetland and open-country birds may be seen here, with resident species including Pygmy Cormorant, Purple Heron, Little Bittern and other herons, all three regularly occurring harrier species, Kingfisher, European Bee-eater and a range of wetland and scrub warblers.

Brent de l'Art
This famous and dramatic canyon, with its layered and sculpted sandstone walls, is a popular tourist destination. Birdwatchers will find much of interest here too, with Grey-headed Woodpecker and Tengmalm's and Pygmy Owls present in the woodlands, and the chance of Dipper and Grey Wagtail by the water and Golden Eagle overhead.

Cima Grappa
This diverse, dramatically beautiful and rugged landscape is great to visit in summer to see classic upland birds like Alpine Chough, Rock Partridge, Alpine Swift, Golden Eagle and Common Rock Thrush.

Ex Polveriera Albignasego
A small LIPU reserve, close to Padua. The circular trail passes through wetland and farmland, and the area is most productive in migration seasons. Nightingale, Cetti's Warbler and Hobby are among the species present.

Ghelpac
This trail, beginning alongside the Ghelpac stream near Fortino, offers easy walking in high-altitude, mostly forested terrain. Special birds of the area include Grey-headed and Black Woodpeckers, Capercaillie, Eagle Owl and Golden Eagle.

Monte Corno
An area of forest and fields where you can encounter Tengmalm's and Eagle Owls, Goshawk, Golden Eagle, and Three-toed and Black Woodpeckers.

Parco dell'Amicizia Tezze sul Brenta
A stretch of floodplain along the Brenta river, this protected area has a 6.5km walking trail through wetlands and farmland with some patches of woodland. It is a great area to look for bitterns, herons and egrets, and you may find Lesser Spotted Woodpecker in the wooded areas.

Parco di San Giuliano Mestre
On the north-western side of the Venetian lagoon, this site provides easy walking and viewing of the sandbanks and lagoon, with farmland and scrub further inland. A great variety of water-birds can be seen including Greater Flamingo, Pygmy Cormorant, Black-necked Grebe, Water Rail and assorted waders and herons.

Parco Nazionale Dolomiti Bellunesi
A very large and diverse protected area in the Dolomites, good for montane species such as Rock Partridge.

Riserva Naturale Integrale Lastoni Selva Pezzi
Upland forest and mountain, notable for raptors, owls and woodpeckers.

Riserva Naturale LIPU di Ca'Roman
A protected stretch of coast, including wildlife-rich intact dune ecosystems. Look for gulls and in summer visiting European Bee-eaters, European Nightjars and Eurasian Scops Owls.

Rubbio
This ski resort is much visited by thrill-seekers, and walkers in summer. You don't have to go too far to find quiet surroundings and a wealth of forest and montane birdlife. Species present include Capercaillie, Nutcracker, Rock Partridge and Golden Eagle.

HOW TO USE THIS BOOK

This pocket guide is designed to be a quick and easy reference book for anyone bird-watching in Italy. The following accounts for individual species are arranged by family and Avibase taxonomic order.

The species text describes the bird's appearance in all plumages likely to be seen in Italy, highlighting key features that will aid identification. Songs and calls are described, followed by information on the bird's habitat preference in Italy, any notable behavioural details, and finally an outline of its distribution, abundance and seasons of occurrence. For those species assigned a conservation status of Near Threatened or Vulnerable by the International Union for Conservation of Nature (IUCN), this is mentioned in brackets at the start of the account. A few technical terms are used, but all are explained in the glossary at the back of the book.

Each account is illustrated with one or more colour photographs, showing the bird as it will be encountered in Italy – so winter visitors are shown in non-breeding plumage, summer visitors in breeding plumage. Passage migrants may be shown in breeding or non-breeding plumage. Where the sexes differ significantly, male and female plumages are shown. The images show wild birds and are chosen to show a clear view in natural viewing conditions. All images depict adult birds unless otherwise stated, and the key below explains the regularly used abbreviations:

Male – ♂
Female – ♀
Juvenile – juv.
Immature – imm.
Breeding – br.
Non-breeding – non-br.

SEASONS

The following terms are used to describe periods of occurrence in Italy for the birds in this book.

Resident Species that can be found year-round. Note, though, that the same species may be resident in some parts of the country but occur only as a visitor in others. For many resident species, the population increases considerably in winter as more birds arrive from northern and eastern Europe.

Summer visitor Species that visit only for the breeding season, migrating south in winter. Present from spring until autumn.

Winter visitor Species that visit only for winter and migrate north to breed. Present from autumn until spring.

Passage migrant Species that neither breed nor winter in Italy, but pass through on their migratory journey, so can usually be seen in spring and again in autumn. Some passage migrants also overwinter or oversummer but in much smaller numbers.

Greylag Goose *Anser anser* 80cm

A large, robust goose with rather uniform grey-brown plumage at rest, marked with narrow pale bars on upperparts and dark barring on flanks. Undertail is white. In flight, shows contrasting light blue-grey forewings, pale grey-brown rump, and grey-brown tail with white at base and tip. Eyes dark, prominent in pale head. Bill large, orange with whitish tip. Legs pale dull pink. Gives loud honking and cackling calls, like farmyard goose (domesticated Greylag). Feral domestic Greylags are usually plumper than wild birds, and often partially or wholly white.

Where to see A passage migrant and winter visitor to lowland wetlands and fields, with small breeding populations in the north-east.

Greater White-fronted Goose *Anser albifrons* 72cm

Smaller than Greylag, with a darker head and neck and less prominent eyes. Adult has a conspicuous white patch around the base of the bill, and variably heavy blackish barring on the belly. Juvenile lacks these field marks, but usually seen alongside adults. Forewing light blue-grey in flight. Bill pink, legs orange-pink. Noisy, with higher-pitched calls than Greylag.

Where to see A localised winter visitor to some lowland wetlands and nearby fields in north-east Italy, usually in flocks with other geese and wildfowl.

Tundra Bean Goose *Anser serrirostris* 72cm

Smaller than Greylag, with more compact build, and relatively short bill and neck. A dusky grey-brown goose with fine paler fringes to upperparts feathers, and white rump-band and undertail. Head very dark, eyes not prominent. Bill black with small orange patch near tip, legs pinkish orange. Wings show little contrast in flight. Gives usual cackling calls, with some very gruff guttural notes.

Where to see A localised winter visitor to some lakes and damp fields in north-east Italy, usually in big flocks with other geese.

Mute Swan *Cygnus olor* 150cm

Very large, graceful, white waterbird with long neck usually held in slight curve. Legs and feet black, bill orange with black markings, pronounced black knob at bill base and bare black skin between eye and bill. Juvenile dull grey, with pink bill, similar black bill and facial markings to adult but lacks bill knob. Often quiet but wings make noticeable swishing sounds in flight.

Where to see Occurs on inland and sheltered coastal waters. Resident and winter visitor mainly in the north.

Whooper Swan *Cygnus cygnus* 150cm

A large, slender swan, similar in size to Mute Swan but less bulky, with slimmer neck. Legs black, bill black with large triangular yellow wedge extending from base to halfway along length. Eye black. Juveniles have light grey plumage, and same face and bill pattern as adults but pale wedge is dull pink. Flocks are noisy, making loud bugling calls.

Where to see Visits damp fields, marshes and other inland wetlands. Rare winter visitor to northern Italy.

Ruddy Shelduck *Tadorna ferruginea* 63cm

A distinctive large duck with rich orange-red plumage, which fades to white on the head in adults. Large white patch on forewing prominent in flight, contrasting with dark flight feathers, with dark green iridescence on secondaries. Male has black neck-ring, most prominent when neck stretched. Bill, eye and legs dark. Gives rather quiet, nasal honking calls.

Where to see Found on marshes and other quiet inland wetlands. A regular but rare migrant and very rare winter visitor.

♀

Shelduck *Tadorna tadorna* 60cm

♂

♀

Large, striking duck. Mostly white with head glossed blackish-green, red-brown breast-band, black belly centre and shoulder-stripes, flight feathers and tail-tip, pale orange undertail. Eyes dark, legs pink, bill pinkish-red (with knob at base in male). Juvenile mostly white with greyish head and upperparts, pink bill. Has various whistling and honking calls. Gregarious.

Where to see Found on flat muddy shorelines and saltmarshes, sometimes lake-shores and grassy fields. Resident in parts of north, also Sicily and Sardinia, more widespread in winter.

Garganey *Spatula querquedula* 39cm

Small, slim and elegant. Male has dark brown head with broad white supercilium, brown neck and breast, grey body with elongated, dark-edged scapulars, spotted tail and undertail. In flight, shows light blue forewing. Female mid-brown with darker mottling, double dark facial stripes, pale spot near bill base. Courting male gives dry rattle, female has soft quack or cackle.

Where to see Nests in undisturbed marshy areas; visits shallow, well-vegetated lakes on migration. Summer visitor, scarce in south; common and widespread passage migrant.

Shoveler *Spatula clypeata* 48cm

Short-necked duck with very long broad bill, usually held tilted downwards. Male has dark green head, white breast, chestnut flanks and belly; shows light blue forewings in flight. Female brown with darker streaking and mottling; in flight shows pale grey forewings and whitish underwings. Calls are hoarse quacks and wheezes. Feeds mainly on surface, with characteristic very flat-backed posture, dabbling with large bill.

Where to see Visits shallow open waters and marshlands. Widespread passage migrant and winter visitor; scarce and localised breeder.

Gadwall *Mareca strepera* 51cm

♂

♀

Slightly smaller and daintier than Mallard. Rather drab – male mostly light grey with jet-black rear end, chestnut patch in wing (often hidden), dark eyes, dark grey bill. Female mottled brown with white belly and underwing, dark bill with orange sides. Both sexes show white speculum in flight. Legs and feet light orange-yellow. Female has Mallard-like quack, male a soft whistle and harsh croak.

Where to see Mainly on well-vegetated lakes. Widespread passage migrant and winter visitor, scarce and localised breeding bird.

Wigeon *Mareca penelope* 46cm

♂ br.

♀

Compact duck with rounded head and small bill. Male has reddish head with creamy crown-stripe, pink breast, grey body with white rear flanks and black tail. Shows large white oval on wing in flight. Female reddish-brown, with white belly and darkish eye-mask. In both sexes eyes dark, bill light grey with black tip. Male gives loud *whee-ooo* whistle, female a growling note. Often grazes in large flocks.

Where to see Damp fields, lakes and marshland. Widespread and locally abundant winter visitor and passage migrant.

Mallard *Anas platyrhynchos* 55cm

Familiar dabbling duck. Male has yellow bill, bottle-green head, narrow white neck-ring, chestnut breast, grey body with black rear (two central feathers tightly up-curled). Female mottled brown, dark eye-stripe, bill brownish-orange. Both sexes have dark eyes and iridescent blue speculum. Female gives loud quack, male soft clucking note. Domestic Mallards frequently occur in the wild – they come in many colours, including black, white, piebald and dilute colours, and may be much smaller or larger than wild Mallards. Males always show the characteristic curled tail feathers.

Where to see All kinds of fresh water, including lakes in town parks. Common, widespread resident.

Pintail *Anas acuta* 55cm (+10cm tail in male)

Slender and elegant. Male has dark brown head and neck, white breast and narrow white line up neck-sides, grey body, white belly, elongated black central tail-feathers. Female cinnamon-brown with unmarked face and long tail, showing narrow whitish wing-bar and white trailing edge to wing in flight. Both sexes have dark eyes, grey bill and legs. Male gives soft purring whistle, female a cawing quack.

Where to see Lakes and marshes with some open water. Widespread winter visitor and passage migrant.

Eurasian Teal *Anas crecca* 36cm

Small, compact, short-necked and relatively small-billed. Male has chestnut head with broad green stripe from eye to nape, edged finely in yellow. Breast buff with dark speckles, undertail yellow with black border. Shows broad white wing-bar in flight. Female is rather plain brown, belly pale, wing-bar narrower. Both sexes have green speculum. Flight fast and wader-like. Male's call a short, high-pitched purring *prroop*, female's a quiet quack. Shy, often reluctant to venture into open water.

Where to see Muddy, well-vegetated lakes and marshland. Localised breeder, common and widespread winter visitor.

Red-crested Pochard *Netta rufina* 55cm

A large duck which dabbles and sometimes dives. Male has an orange-red head with squarish, bouffant crown, black breast and rear end, brown back and whitish flanks; bill and eye red. Female dull brown with contrasting paler grey cheeks, chin and throat, faint paler barring on body, dark bill with pink patch, dark eyes. In flight male shows mostly white flight feathers with dark tips; female's are pale brown.

Where to see Lakes and estuarine wetlands. A patchily distributed resident, mainly in north Italy and Sardinia, more widespread as passage migrant.

Pochard *Aythya ferina* 45cm

(Vulnerable) Compact, with peaked crown and sloping forehead. Male has red-orange head, red eyes, black breast and rear, and light silvery-grey wings and body. Female light grey-brown with dark eyes, diffusely barred body and faint pale 'spectacles', but rather variable. Bill dark with wide grey band. Female gives purring call in flight. Dives, and feeds at surface.

Where to see Breeds on well-vegetated lakes and marshland with some deeper water; winters on any open waters. Localised resident, widespread and locally abundant in winter.

Ferruginous Duck *Aythya nyroca* 40cm

(Near Threatened) Small, dark with peaked head. Male rich reddish-brown with white undertail and white eyes. Female browner, eyes dark. In flight shows wide white wing-bar and white belly. Bill grey in both sexes with black tip and paler band near tip. Note that some female Tufted Ducks show a white undertail. Female gives a purring flight call; male whistles and clucks in courtship. Shy when breeding.

Where to see Breeds on lush marshy lakes. Very localised; more widespread on migration.

♂

♀

Tufted Duck *Aythya fuligula* 44cm

Smallish, with crest at back of head. Male glossy black with white flanks and long drooping crest, female warm dark brown with paler flanks; may have white 'blaze' around bill base and/or white undertail. Both sexes have yellow eyes and show white wing-bars in flight. Juvenile has darker eyes. Bill grey with broad black tip. Female has growling call, courting male a fast chatter. Dives for food.

Where to see Prefers lakes with some deep water. Localised resident but widespread winter visitor.

Scaup *Aythya marila* 47cm

Sleek, with rounded head. Male has black breast and rear, green-glossed black head, and finely barred silver-grey body plumage (slightly darker on back than flanks). Female brown with large white patch around bill base, and pale cheek patches. Both sexes have yellow eyes, pale grey bill with small black tip, and show wide white wing-bar in flight. Gives low, grating calls.

Where to see Sheltered bays and estuaries, deep inland lakes. A scarce winter visitor, most likely in north-east Italy.

Common Eider *Somateria mollissima* 65cm

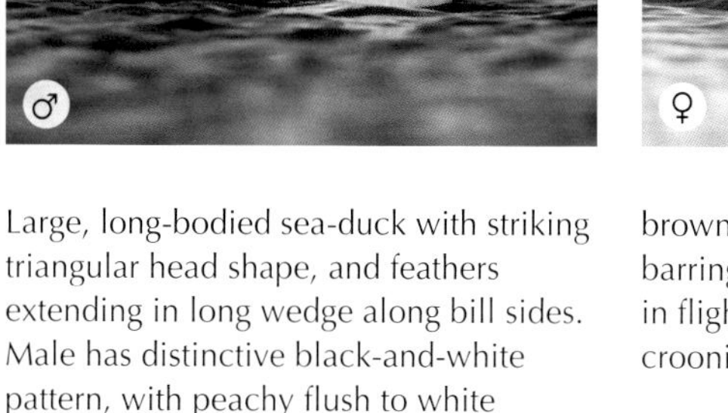

Large, long-bodied sea-duck with striking triangular head shape, and feathers extending in long wedge along bill sides. Male has distinctive black-and-white pattern, with peachy flush to white breast, yellow-based grey bill, and dull green patches on neck sides. Young and eclipse males mostly black. Female brown or grey-brown with fine darker barring, shows some white on underwing in flight. Dives for food; males give crooning calls in courtship.

Where to see Sheltered bays and estuaries. Scarce winter visitor on northern coasts. A handful of pairs breed.

Velvet Scoter *Melanitta fusca* 55cm

Large sea-duck with pointed tail. Male black with white flash around pale eye, yellow sides to black bill. Female dark brown with two paler face patches (in front of and behind eye). Both sexes have white wing-bar (visible in flight and when birds resting on water stretch and flap). Makes long dives to feed, usually seen in 'rafts' resting well offshore. Will mix with Common Scoters.

Where to see At sea, in sheltered waters. Uncommon winter visitor to northern coasts and large inland lakes.

Common Scoter *Melanitta nigra* 50cm

Slightly smaller than Velvet Scoter. Male entirely black with narrow yellow centre to black bill; eye dark, bill base has swollen knob. Female dark grey-brown with paler dusky greyish cheeks and neck. Lacks pale wing-bar. Forms flocks or rafts on the sea; they dive in synchrony when feeding. May be seen flying in straggly lines low over the water.

Where to see At sea, in sheltered waters, occasionally on inland lakes. Uncommon winter visitor, most likely off northern coasts.

Goldeneye *Bucephala clangula* 46cm

Smallish, compact diving duck with peaked head shape. Male predominantly white with black markings on wings and tail, head black with green gloss, white circular marking at bill base. Female mainly grey, faintly barred darker, with whitish neck and chestnut-brown head. Both sexes have yellow eyes, grey bill (with yellow marking near tip in female), and show much white in wings in flight. Dives frequently.

Where to see Deep lakes and sheltered coastal waters. Winter visitor, most likely in north where it can be locally common.

Goosander *Mergus merganser* 63cm

Large, long-bodied and streamlined diving duck with long, narrow, serrated bill for fish-catching. Male mainly white with some black upperside markings, green-glossed black head. Female mainly grey with slightly shaggy dark reddish-brown head (neatly demarcated from pale neck) and white chin. Both sexes have dark eyes, reddish bill, extensive white in wings noticeable in flight. Gives croaking and rattling calls.

Where to see Breeds locally in north Italy, mainly on rivers and lakes. At other times visits deep inland waters and coasts.

Red-breasted Merganser *Mergus serrator* 55cm

Slender, with shaggy crest. Male has dark green-glossed head, white neck-ring, reddish breast, black-and-white upperparts and grey underparts. Female and eclipse male drabber with reddish-brown head. Both sexes have shaggy crests, reddish eyes and show much white in wings in flight. Bill red, long, narrow with hooked tip. Call a low, grating *kraak kraak*. Sits low in water and makes frequent lengthy dives.

Where to see Occurs offshore on sheltered coasts, sometimes in groups. Passage migrant and winter visitor; some oversummer.

Hazel Grouse *Tetrastes bonasia* 37cm

A small, long-tailed grouse with short, peaked crest (sometimes held flatter). Plumage finely and intricately marked with brown, grey and black streaks, scalloping and mottling; has blackish tail-tip, unmarked grey rump, and white 'shoulder straps'. Male has black throat, female's throat speckled grey and white. Shy and keeps in cover, relying on camouflage, but takes rapid flight through the trees as soon as startled. Has short, fluting alarm call, and male's song is series of thin, trilling, high-pitched notes, recalling a much smaller bird.

Where to see Found in dense pine and birch forest in alpine regions; resident.

♂

♀

Capercaillie *Tetrao urogallus* 60–88cm

A very large, imposing grouse, recalling turkey in build, with long, heavy tail. Male blackish, shading to dark brown on back, with some paler markings on flanks, undertail and tail. Small red wattle above eye. Bill strong, hooked, pale horn in colour. Female smaller but still a substantial bird; finely patterned in grey-brown and black, with rusty tinge on breast and throat, bill dark. Shy and quick to flush, often perching in trees. Quiet except when displaying at communal leks, when males strut with fanned tails, giving loud croaks and grunts.

Where to see Resident in mature pine forest in alpine regions.

♀

♂

Black Grouse *Lyrurus tetrix* 40–58cm

Smaller, daintier than Capercaillie. Male black with white undertail and wing-bars. Long lyre-shaped tail, red wattles over eyes, bill small and dark. Female smaller, barred and mottled grey-brown; shorter blunt-ended tail. Differs from much larger female Capercaillie in lacking reddish tint on breast; also shows narrow white wing-bars in flight. Males display communally at leks, leaping, fighting and posturing, flaring white undertail, giving cackling and bubbling calls.

Where to see Pine and birch forest, moorland, heaths and bogs. Resident in Alps.

Grey Partridge *Perdix perdix* 30cm

Small, rotund and stocky gamebird with short tail. Predominantly grey and sandy-brown in muted pattern; face orange and flanks strongly barred reddish-brown. Eyes dark, bill grey, legs dull pinkish. Male more colourful with bold dark horseshoe-marking on belly, this reduced in female. Juvenile brown with darker and paler streaks and barring. Male has explosive, ringing two-note song. Gregarious, forages discreetly on ground and usually runs when alarmed.

Where to see Farmland, woodland edges and other well-vegetated open habitats, from plains to hills. Now rare and very local in central and northern Italy.

Ptarmigan *Lagopus muta* 33cm

♀ br.

Elegant, smallish grouse. In summer, grey-brown above with white wings, belly and (fully feathered) legs – male has black lores and red 'combs' above eyes. In winter entirely white, except for tail, which remains black in all plumages. During moult, upperside plumage patchy grey and white. Gives snorting or snoring calls. Gregarious, especially in winter and sometimes confiding, moving slowly across rocky ground in search of food. Fast direct flight.

Where to see High snowy mountains, only in Alps in Italy.

♂ non-br.

Common Pheasant *Phasianus colchicus* 60–80cm

A well-known gamebird. Adult males are unmistakeable, with dark green head ornamented with ear-tufts and scarlet facial wattles. Some have white neck-ring. Body plumage rich reds, blue-greys and browns with dark and white markings, tail very long with dark barring. Female plain mottled brown with long tail, but short-tailed young birds could be confused with other gamebirds. Melanistic and leucistic (white) forms may occur. Call a loud, crowing two-note cluck; male's song a longer series, accompanied with noisy wing-fluttering.

Where to see All kinds of countryside habitats with good ground cover. Introduced to Italy for shooting; common, widespread resident.

Common Quail *Coturnix coturnix* 17cm

Tiny, plump, short-necked and short-tailed, but with long wings and fast flight. Plumage light grey-brown with broad pale and dark streaks, most prominently on flanks. Has pale supercilium and crown-stripe; male has black centre to throat. Very shy and difficult to see, most often detected by male's song given at dawn and dusk – a repeated, rhythmic three-note whistle with longer space between notes one and two; '*wet... my-lips*'.

Where to see Open, well-vegetated grasslands and dense crop-fields. Widespread summer visitor.

Barbary Partridge *Alectoris barbara* 33cm

Resembles Red-legged Partridge but has grey rather than white chin and supercilium, giving plainer appearance. Body plumage grey-brown with boldly barred flanks, orange-brown belly and undertail, broad blackish crown-stripe, reddish eye-stripe, grey chin and throat bordered with streaky dark reddish 'neck-shawl'. Bill and legs red. Call a series of short notes with fast, regular rhythm, also raucous three-note call like that of Red-legged Partridge.

Where to see Found in a variety of habitats, especially more open and rocky terrain. Resident and fairly common on Sardinia – one of only a handful of European populations of this primarily North African species.

Red-legged Partridge *Alectoris rufa* 33cm

Very like Rock Partridge, but has smaller white chin patch, and black collar extends onto greyish neck as streaky dark 'neck-shawl'. Also has white supercilium, less bold flank streaks (black and light grey rather than black and white). Like other partridges is reluctant to fly, preferring to escape danger on foot. Flies fast with shallow wingbeats, and in flight shows dark red-brown tail-sides, contrasting with blue-grey of lower back and rump. Most familiar call is harsh three-note *kuh chuh-chuh* but will give notes in longer series at times.

Where to see Lives in rocky terrain, farmland, field edges; only in north-west Italy.

Rock Partridge *Alectoris graeca* 34cm

(Near Threatened) Stout, colourful partridge. Blue-grey, shading to brown on back and yellowish-brown on belly. Broad black-and-white barring on flanks. Chin and throat white, boldly bordered with black (less prominently in Sicilian subspecies *whitakeri*). Bill, legs and eye-rings scarlet. Song a rapid series of hoarse notes. Gregarious and sometimes conspicuous – will perch on prominent rocks.

Where to see Rocky and mountainous areas with some vegetation, mainly 1,000–2,000 metres (but avoids north-facing slopes). Resident on high Alps, Apennines and southern mountains.

Greater Flamingo *Phoenicopterus roseus* 133cm

Unmistakeable. Tall with very long legs and very long, slender neck. Large, deep-based, steeply downcurved bill. At rest looks mainly whitish-pink, with legs and bill darker, bill tipped black, neck held in graceful 'S' curve. In flight holds neck and legs fully outstretched; black flight feathers and bright pinkish-red underwings and forewings. Juvenile drab grey-brown with dark legs, gradually whitening with maturity. Gives goose-like cackles in flight. Feeding flocks move together in close proximity, heads lowered upside-down into water, giving regular low-pitched murmur.

Where to see Found at shallow lagoons, sea bays and salt lakes. Gregarious but localised colonial breeder; more widespread in winter and on migration.

Little Grebe *Tachybaptus ruficollis* 26cm

Small, compact waterbird. Short, pointed bill. Looks tailless and fluffy-bottomed. Breeding plumage blackish with dark chestnut cheeks, neck and flanks, whitish rear, circle of bare yellow skin around gape. Bill and eyes dark. Juvenile and non-breeding plumage paler, with brown upperparts and warm buff underparts, bill paler, gape marking less prominent. Chick dark with paler stripes; juvenile may show faint stripes. Gives loud whinnying chatter when breeding. Dives frequently when feeding.

Where to see All inland waters; common, widespread resident.

Slavonian Grebe *Podiceps auritus* 35cm

A small but elegant grebe with shortish, straight bill and flat-topped head. Bill shape and head shape help distinguish it from Black-necked Grebe. In breeding plumage (unlikely to be observed in Italy) colourful in reddish, black and yellow. Non-breeding birds look monochrome with dark grey upperparts and whitish below, blackish crown contrasting with white cheeks, and prominent red eyes. Makes long, frequent dives.

Where to see Visitor to sheltered shorelines and estuary mouths, scarcer inland. Uncommon winter visitor.

Red-necked Grebe *Podiceps grisegena* 43cm

Somewhat smaller and darker than Great Crested Grebe. In breeding plumage (unlikely to be observed in Italy) has reddish neck and flanks, dark crown and whitish cheeks. Bill dark with yellow base. In non-breeding plumage rather dusky dark grey on upperside. Crown blackish, reaching below eyes, with paler cheeks, chin and lower breast; does not show the all-white foreneck of Great Crested.

Where to see Offshore in sheltered waters, sometimes inland. A scarce winter visitor, mainly to northern areas.

Great Crested Grebe *Podiceps cristatus* 49cm

Slender, elegant. Upperparts brown, flanks lighter buff-grey, otherwise mostly white with black crown and lores. In breeding plumage has crest and chestnut, black-edged facial ruff, dark bill. In winter lacks ruff and tufts, bill becomes pink. Chick striped black and white; juvenile shows traces of stripes on face until early winter. Noisy croaks and rattling calls in breeding season, especially during courtship. Dives often.

Where to see Nests on inland waters; in winter also at sea in sheltered waters. Common, widespread resident.

Black-necked Grebe *Podiceps nigricollis* 31cm

Small, dainty. Markedly peaked crown, slim bill with upward tilt. In breeding plumage has black head, neck and breast, reddish flanks, yellow tufts behind bright red eyes. In winter blackish-grey above, whitish below, with dusky collar and whitish rear end. Chick grey with black-and-white striped head, juvenile like adult winter but browner with duller eyes. Gives *pu-iih* call when breeding.

Where to see Breeds in loose colonies on well-vegetated inland waters. Winters in estuaries, bays and lakes. Localised breeding bird; widespread in winter.

Rock Dove/Feral Pigeon *Columba livia* 33cm

Rock Dove is the ancestor of all domestic pigeons and their feral 'street pigeon' descendants. A stocky, medium-sized pigeon, usually with orange eyes, red legs and black bill with white cere. Plumage extremely variable – majority are grey with dark-barred or chequered wings, and some green and violet neck iridescence. Shows white underwings in flight. Other variants include black, entirely white, white-patched and red-brown. Gregarious, gives various cooing calls.

Where to see Feral Pigeon: a widespread resident in farmland, villages and urban areas. Rock Dove: rare and very local in craggy coasts.

Stock Dove *Columba oenas* 31cm

Smaller than Woodpigeon and Feral Pigeon, may mix with both species. Rather plain grey, lacking contrast, with short dark double wing-bars. Eyes beady and black, bill yellow with pink base, has some green and purple iridescence on neck. In flight shows grey rather than white underwings. Usually rather shy, may be seen foraging on ground or resting in trees, often close to nest hole. Territorial call a single, emphatic low-pitched *coo*.

Where to see Woodland edges, farmland, parks. Rare and local resident mainly in north-west Italy. More common and widespread in winter.

Woodpigeon *Columba palumbus* 40cm

Plump, small-headed with waddling walk. Mainly light grey, pinker on breast and browner on wings. White neck patch and wing edge. In flight shows broad white wing-band. Eye whitish, bill yellow with pink base, legs pink. Juvenile lacks white neck patch, has darker eyes and bill. Feeds on ground and in trees. Performs display flight with wing-clapping ascent, gliding descent. Song a loud five-note cooing phrase. Gregarious in winter.

Where to see Woodland, parks, gardens, farmland. Common and widespread resident.

Turtle Dove *Streptopelia turtur* 26cm

(Vulnerable) Smallish, long-tailed, colourful dove. Head grey, shading to pinkish-grey on breast. Black-and-white striped patch on neck. Back and wing feathers dark with bright orange-brown fringes, giving 'tortoiseshell' pattern. In flight shows distinctive tail pattern – underside white with broad black base, upperparts dark with white edges; underwings grey. Bill small and slim, prominent red eye-rings, legs reddish. Juvenile patterned like adult but paler and drabber, lacks striped neck patch. Song a soft, soothing purr or croon, often given from overhead wire or other prominent perch.

Where to see Favours woodland and farmland with scattered trees and copses. A common but declining summer visitor and passage migrant.

Collared Dove *Streptopelia decaocto* 31cm

Larger than Turtle Dove. Plumage light pinkish-fawn with narrow, white-edged, black half-collar on neck (absent in juvenile). In flight shows dark wingtips, pale greyish band across forewings, pale tail-sides, undertail white with dark base. Eyes dark reddish, bill small and dark, legs pink. Song recalls Woodpigeon but comprises three notes, with stress on middle note. Gregarious in winter, can be very confiding.

Where to see Lowlands where food supplies are abundant, especially on farmland and around settlements. A common resident.

Little Bustard *Tetrax tetrax* 44cm

A pheasant-sized but long-necked and long-legged bird, with small head and generally sandy-brown coloration with white underparts. Eyes reddish, sturdy legs and small bill are greyish. In flight, shows white flight feathers with black tips. Male has black neck and upper breast (puffed up in display), broken by two white stripes, and greyish face. Gregarious in winter. Generally shy and unobtrusive but displaying males are noisy and showy.

♂

Where to see Occurs on grassland and farmland. A scarce resident in parts of Sardinia.

European Nightjar *Caprimulgus europaeus* 26cm

Nocturnal, with long tail and wings. Intricately mottled, speckled and barred in grey and brown, with whitish 'tramlines' on back, pale markings on cheeks and chin-sides. Male in flight shows white tail-corners and white patches near wingtips. Rests by day on ground or perching along branch, relying on camouflage. At night, hawks for moths in agile, silent flight. Male gives continuous churring song from dusk; also wing-clapping display flight.

Where to see Breeds on heathland, moors, bogs and forest clearings. Widespread summer visitor.

Great Spotted Cuckoo *Clamator glandarius* 39cm

Large, striking, long-tailed bird. Adult dark grey above with grey crest, whitish below with yellow wash on throat and breast-sides, wings marked with prominent white spots. Bill, eyes and legs dark. Juvenile more brown-toned with dark golden-brown flight feathers and black crown. Has loud chattering call. A brood parasite, laying eggs in other birds' nests (most often Magpie) – the cuckoo chicks are raised alongside the foster parents' own offspring.

Where to see Lightly wooded and other semi-open landscapes, usually not far from coasts. Scarce and sparsely distributed summer visitor.

Common Cuckoo *Cuculus canorus* 34cm

A hawk-like bird with long tail and long pointed wings, short yellow legs. Distinctive perched posture with wings drooped. Upperparts grey (tinged brown in female), belly white with fine black bars; female also occurs in rare rufous ('hepatic') morph. Juvenile browner, barred all over. Bill small, slightly downcurved. Has yellow eye-rings, light brown eyes. Male's song is two-note hollow *cuk-oo*, female has a fast mellow bubbling trill. A brood parasite of small birds such as Robins and Reed Warblers. The cuckoo chick ejects host's eggs/chicks and is reared alone by host parents.

Where to see Common summer visitor, adults departing in early summer, juveniles in autumn.

Alpine Swift *Tachymarptis melba* 21cm

Largest swift; sturdy and powerful in flight. Wings long, sickle-shaped, tail forked. Upperparts dull mid-brown, underparts similar but with clean-cut white throat patch and belly – though throat patch hard to see at distance. Flies with slower, less flickering wingbeats than smaller swift species. Loud, long, rapid twittering call, falling in pace and pitch. Flies in groups around breeding grounds and may forage long distances from nest, catching insects in flight and (when feeding young) collecting bolus of food in throat pouch, forming obvious swelling to chin.

Where to see Nests in rocky crevices and buildings in highlands and lowlands. A scarce but widespread summer visitor and passage migrant.

Common Swift *Apus apus* 17cm

Blackish-brown with small paler throat patch. Juvenile slightly paler overall with Pallid-like scaly body plumage. Agile aerial hunter of insects, trapping them in wide, hair-fringed mouth. Has fast flickering flight with rapid twists and turns. Flies high in fine weather but in bad weather will skim ground. Flocks give thin drawn-out screaming calls (with birds in nests answering in kind).

Where to see Nests mainly in crevices in buildings and hunts over open country, wetlands and urban environments. Very common, widespread summer visitor.

Pallid Swift *Apus pallidus* 17cm

Mid-brown swift with large, diffuse pale throat patch, small dark 'mask' around eyes, and pale fringes to body feathers, giving scaly appearance. Slightly broader-winged with slower flight than Common Swift, with lower-pitched screaming call; wings show more contrast between dark coverts and paler secondaries; when seen from above shows contrast between dark rump and pale inner secondaries.

Where to see Nests in crevices on buildings, most common near coast. Summer visitor, much less abundant than Common Swift.

Water Rail *Rallus aquaticus* 24cm

A shy bird of well-vegetated wetlands. Warm brown above with black streaks, plain lead-grey face and underparts, with fine black-and-white barring on flanks. Stout build with longish neck, long red bill, short, pointed tail often cocked to reveal white underside; round-winged in flight. Legs pink, longish with very long, slim toes. Juvenile paler and browner with yellowish bill and legs, lacks lead-grey tones, pale face with darkish eye-stripe. Has loud squealing piglet-like call. Forages discreetly in reeds and sedges at water's edge, and sometimes swims. May forage more in open in winter.

Where to see Lowland lakes and marshland. A widespread resident.

Corncrake *Crex crex* 28cm

A dry-country crake, somewhat recalling a partridge. Plumage muted but complex, with mainly grey head and neck, rufous barring on flanks, and light brown upperparts with strong dark streaks. In flight, uppersides of wings look very rufous. Moves discreetly through tall vegetation, when flushed flies short distances with legs dangling. Male's territorial song is constant series of slow-paced, rasping *crex-crex* notes, delivered in a tall upright posture.

Where to see Scarce summer visitor to lush meadows and farmland in the Alps, mainly in the eastern section.

Spotted Crake *Porzana porzana* 21cm

A very shy bird of rough wet grassland. A little smaller than Water Rail, longer-winged, with much shorter, thick-based bill. Plumage basically grey on underparts, brown on upperparts, heavily marked with white spotting below (becoming barring on flanks), and black spots above. Front of face blackish, undertail unmarked whitish-buff. Bill yellow-grey with red spot at base, legs dull green. Juvenile paler, with duller bill. Male's territorial 'song' a distinctive upslurred whistle, recalling whip-crack, most often heard at night. Stays in cover, slips out of view when disturbed, reluctant to fly.

Where to see Lowland lakes and marshland. Widespread on migration, rare and very localised breeder.

Moorhen *Gallinula chloropus* 29cm

Familiar, conspicuous waterbird. Plumage blackish, greyer on underparts, brownish on wings. Has white undertail and flank-stripes. Red frontal shield, bill red with yellow tip. Legs dull green. Juvenile has similar white markings but otherwise grey-brown, paler on underparts. Very vocal, with various short sneezing notes and longer liquid calls. Swims with tail raised, bobbing head, feeds at surface. Flicks tail constantly when walking on land. Often perches in trees.

Where to see Found in all kinds of well-vegetated wetlands. Common resident.

Coot *Fulica atra* 39cm

Stout, almost tailless waterbird. Adult sooty grey-black. Wing shows narrow white trailing edge in flight. Eyes dark red, bill and frontal shield white. Legs and lobed toes grey. Juvenile greyer, whitish on underparts. Downy chick black with bald crown and orange filoplumes on head (lacking in otherwise similar Moorhen chick). Dives for food, grazes on lake shores. Gregarious and vocal; gives various croaks and sharp calls.

Where to see Found on reservoirs, lakes and slow-flowing, large rivers. Common resident.

Purple Swamphen *Porphyrio porphyrio* 48cm

Large and long-legged relative of Moorhen. Blackish plumage with violet-blue gloss particularly on head, white undertail. Has large red frontal shield and powerful, thick bill, red eyes. Legs pinkish-red. Juvenile ashy-grey with darker bill and shield. Swims with rear end very elevated, more often seen walking or wading along shorelines, flicking its tail. Has varied calls, similar to those of Moorhen but lower-pitched.

Where to see Found in marshy wetlands with lush marginal vegetation. Occurs on Sardinia and Sicily.

Little Crake *Zapornia parva* 18cm

♂

♀

Small, fairly long-legged. Male patterned like Water Rail with grey face and underparts, brown upperparts but with plain grey, unbarred flanks and barred undertail. Female has grey only on face, underparts otherwise pale buff; juvenile similar but with brown-barred flanks. Bill small, yellow with red base. Legs greenish. Male's nocturnal song is an accelerating series of short barking *kua* notes. Shy. Sometimes swims; climbs in low reed stems.

Where to see Reedy and marshy areas. Scarce summer visitor to northern Italy, and widespread passage migrant.

Common Crane *Grus grus* 110cm

Very large, stately bird, with long legs and neck, elongated tertial feathers creating appearance of large shaggy 'tail' when perched. Plumage grey, back mottled with brown in breeding season. Rear of neck white, throat and crown black with small red central crown patch. Eyes red, shortish bill horn-coloured, legs grey. Juvenile has plain buffish head and lacks shaggy rear end. Flies with neck and legs outstretched, shows black flight feathers – small white 'landing light' marking at wing-bend. Loud bugling call in flight.

Where to see A scarce and local winter visitor but common passage migrant, most often seen flying in long skeins; flocks also feed on fields.

Stone-curlew *Burhinus oedicnemus* 41cm

Distinctive large-headed, long-tailed wader of dry open countryside. Plumage light brown with heavy, coarse dark streaking, pale stripes above and below eye, pale dark-edged wing-band, white belly and pale underwings with dark edge. Juvenile less boldly marked. In flight shows black flight feathers with white patches. Bill short, dark with yellow base. Eyes golden, very large. Legs yellow. Call a Curlew-like *kur-leee*, song (given at night) an extended series of wailing notes. Relies on camouflage to avoid detection. Mainly nocturnal.

Where to see Forages and nests on all kinds of open, sparsely vegetated ground. A fairly widespread but scarce summer visitor.

Black-winged Stilt *Himantopus himantopus* 35cm

Unmistakeable slim wader with long pointed wings and preposterously long, bright candy-pink legs. White head (smudged blackish in female) and body, black back and wings, variable black or grey markings on head and hind-neck (some pure white). Bill mid-length, straight, black, very fine. Eyes dark red. Juvenile similar but dark parts brownish, scaly-looking; dusky wash on face and hind-neck. Loud grating flight call, given very persistently at breeding colonies when mobbing intruders. Wades in deep water, picking food from surface.

juv.

Where to see Nests on lake islands or shores, mainly near coast; migrants visit coastal marshland and lagoons. Widespread summer visitor and passage migrant, rare and local in winter.

♀

Avocet *Recurvirostra avosetta* 44cm

Unmistakeable, elegant wader with unique upswept bill. Plumage mainly white, with black crown and hind-neck, black wing markings and primaries. Bill thin, black, with strong up-curve; legs long, greyish. Juvenile like adult but dark areas paler. Chick grey with short, straight bill. Call a series of loud, ringing notes. Feeds very actively, sweeping bill side to side; wades in deep water, sometimes swims and even up-ends. Sociable, feeds and flies in flocks. At breeding colonies, all birds rise to fiercely mob any intruder.

Where to see Nests colonially on lagoon islands and shores. Breeds sparsely in northern Italy, also Sardinia and Sicily. Widespread at coast and inland on passage and in winter.

Oystercatcher *Haematopus ostralegus* 42cm

non-br.

(Near Threatened) Distinctive, conspicuous and noisy black-and-white shorebird. Robust, stocky build. Develops narrow white throat-collar in winter. In flight shows broad white wing-bar, tail base, rump and lower back. Bill long, straight and bright orange (with dusky tip in winter), stout legs pink. Juvenile sooty-grey above, bare parts duller. Loud ringing, piping calls. Forages for molluscs on seashore and rocks.

Where to see Rocky or sandy beaches, sometimes lake or river shores inland. Scarce summer visitor in north-east, passage migrant or winter visitor elsewhere.

Grey Plover *Pluvialis squatarola* 27cm

imm.

Rotund and robust plover. In winter uniform dull mottled grey; juvenile similar but more brown-toned. Breeding adult (rare in Italy) has black face and underparts. Bill black, short and stout; large eyes dark, legs dark grey. In flight shows white wing-bar and black 'armpits'. Call a plaintive three-note whistle, *peee-uu-ee*. Runs and pauses while foraging. Not especially gregarious but mixes with other shoreline waders.

Where to see A shorebird, uncommon inland. Passage migrant and winter visitor.

Golden Plover *Pluvialis apricaria* 26cm

Smaller, slimmer than Grey Plover with finer bill. In winter, uniform dull yellowish-grey with fine speckling; has blackish underside in breeding plumage. Legs, bill and eyes dark. In flight shows hint of narrow white wing-bar; 'armpit' area white. Flight call single, slightly downslurred whistle. Agile on the wing, flocks seeming to sparkle white as they turn together.

Where to see Damp grassland, marshes, lake shores; much less frequent on beaches than Grey Plover. Widespread passage migrant and winter visitor to Italy.

Lapwing *Vanellus vanellus* 30cm

(Near Threatened) Distinctive shorebird. Mainly white below, black above with green and violet gloss. Long, fine black crest. Legs pinkish, bill black. Undertail orange. Juvenile duller with shorter crest. In flight shows white underside to inner wings, white at primary tips. Wings distinctively round-ended, broad at 'hand'; agile flyer with tumbling territorial flight. Call a high, excitable two-note *pee-iich*; song similar, interspersed with wheezing, grating sounds.

Where to see Grassland and open muddy ground. Common winter visitor and passage migrant, resident in north.

Kentish Plover *Charadrius alexandrinus* 16cm

Smaller and more lightly marked than Little Ringed Plover, with shorter rear end and proportionately large head. Male brown above and white below with narrow black broken breast-band, thin black eye-stripe, white supercilium, rufous tint on rear crown. Shows broad white wing-bar and rump-sides in flight. Bill slender and dark, legs dark. Female, non-breeding and juveniles paler and duller. Call a soft rolling purred whistle or an upslurred *tew-it*. Behaviour like other ringed plovers; often confiding.

Where to see Nests on muddy or gravelly bare ground near water and close to coast (including on beaches). Scarce and declining resident, also passage migrant.

Ringed Plover *Charadrius hiaticula* 18cm

Plump, rounded with bold markings. Upperparts light brown, underparts white, with broad black breast-band; black eye-stripes and forehead-band, no obvious eye-rings. Bill orange with black tip, legs orange. Black markings become dark brown and bare parts duller in winter and juvenile plumage. Shows narrow white wing-bar in flight. Call an upslurred whistle: *tuu-ip*. Runs and pauses when looking food.

Where to see Sandy and stony beaches; sometimes lake shores and riversides. Usually in small flocks. Winter visitor and passage migrant in Italy; widespread.

Little Ringed Plover *Charadrius dubius* 17cm

Slimmer than Ringed Plover, with clearly longer legs and rear end. Breast-band narrower, has white line between black forehead-band and brown crown. Conspicuous yellow eye-rings. Legs dull pink, bill dark and slim. No obvious wing-bar in flight. Call a short downslurred *peeoo*. Forages on shoreline; in breeding season very active and vocal, chasing rivals; runs to and fro with flank feathers puffed out.

Where to see Nests on gravel shores of lake islands and riversides. Common and widespread summer visitor and passage migrant.

Dotterel *Charadrius morinellus* 21cm

Smallish plover, with distinctive broad pale supercilium and narrow pale breast-band in all plumages. Breeding plumage quite colourful with reddish underside shading to black on belly, undertail white. Upperside grey-brown with paler fringes to wing feathers, crown black and lower face whitish. Non-breeding adult and juvenile plainer and paler. Eyes large and dark, bill blackish, legs dull yellow. Approachable, has ringing *weet-weet* call.

Where to see A montane breeder. Extremely rare in Alps where one or two pairs may breed irregularly. Small groups stop off on bare hilltops and mountains on autumn migration.

Whimbrel *Numenius phaeopus* 41cm

Smaller than Curlew. Plumage brown with fine dark streaks on underparts, mottling on upperparts. Dark eye-stripe, pale supercilium and dark crown. Bill longish with distinct downward curve after halfway point. Bill and legs dull grey. In flight shows white wedge on rump, all-dark wings with blackish primaries. Fast rippling flight call of seven or so notes on same pitch.

Where to see Muddy seashores and around lakes, singly or in small groups. Passage migrant, mainly in north, more frequent in spring than autumn.

Curlew *Numenius arquata* 54cm

(Near Threatened) Larger, longer-legged and longer-billed than Whimbrel. Streaked and mottled brown, without Whimbrel's bold face stripes; eye prominent in rather plain head. Bill smoothly downcurved; can be very long, with pinkish base. Legs grey. White rump wedge in flight, wings look greyish with dark tips. Call melodious upslurred *coor-leee*. Usually in small groups or alone, foraging at stately pace.

Where to see Muddy shores and fields, coastal and inland. Widespread passage migrant and winter visitor; extremely rare and localised breeder in north-west.

Bar-tailed Godwit *Limosa lapponica* 36cm

(Near Threatened) Large, long-legged, with long slightly upswept bill. In breeding plumage deep red on head and underparts, dark eye-stripe and crown, mottled dark grey-brown on upperparts; bill dark. In winter pale grey-brown with mottled upperparts, long supercilium. Juvenile similar but browner. Legs dark. Shows barred tail and white rump wedge in flight; no wing-bar; pattern recalls Curlew. Call a sharp *kwee kwee*.

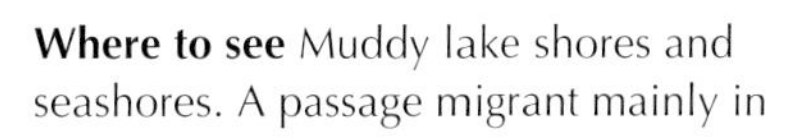

Where to see Muddy lake shores and seashores. A passage migrant mainly in northern Italy; rare and local in winter.

Black-tailed Godwit *Limosa limosa* 39cm

(Near Threatened) Longer-legged than Bar-tailed Godwit. In breeding plumage orange-red on head and breast, belly white with black barring, darkish eye-stripe and crown, lightly mottled grey-brown on upperparts. In winter pale grey-brown, whiter below, short supercilium. Juvenile has peachy face/breast. In flight shows broad white wing-bar, black tail and square white rump patch. Call a shrill three-note whistle. Gregarious.

Where to see Most frequent in freshwater habitats. Widespread passage migrant, scarce winter visitor and rare and localised breeder in north-west.

Turnstone *Arenaria interpres* 22cm

Distinctive stout, smallish shorebird. Has white underparts and in breeding plumage (rare in Italy) boldly marked black and bright chestnut upperparts, fading to dark mottled grey-brown in winter, with blackish breast-band. Legs short, orange. Bill short, wedge-shaped, dark. In flight appears dark with white tail-band, back, wing-bar and shoulder-braces. Flight call a rattling chuckle. Explores strandline and rocks, often in small flocks; can be very approachable.

Where to see Stony and rocky seashores. Passage migrant and winter visitor, mainly on northern coasts.

Knot *Calidris canutus* 24cm

non-br.

Stocky, fairly small sandpiper with shortish legs, larger and chubbier than Dunlin. In breeding plumage (rare in Italy) has red face and underside, mottled grey-and-brown upperparts. In winter light silver-grey with faint streaking on underparts; juvenile similar but slightly browner. Bill mid-length, black, straight, legs dull yellowish. In flight shows narrow white wing-bar and faintly paler rump. May form large flocks, mixes with Dunlins and other shoreline waders.

Where to see Most frequent on seashores and estuaries. Passage migrant, scarce in winter and never in large flocks as in northern Europe.

Ruff *Calidris pugnax* 24–30cm

juv.

♂ non-br.

Large, long-legged with relatively short bill and small head. Male (larger than female) in breeding plumage sports large head-ruff and ear-tufts which may be buff, brown, rufous, black or white. Non-breeding male and female grey-brown with scaly back pattern. In flight shows narrow white wing-bar and rump-sides. Legs yellowish or orange, bill dark with orange base. Juvenile has dark bill, duller legs, warm buff plumage tone. Rarely calls.

Where to see Marshland, wet grassland, lake and lagoon shores. Widespread, sometimes abundant, passage migrant; rare and localised winter visitor.

Curlew Sandpiper *Calidris ferruginea* 20cm

juv.

br.

(Near Threatened) Smallish, elegant and long-legged. In breeding plumage dark brick-red below, mottled blackish and grey above. In winter light silver-grey with clean white belly, white supercilium. Juvenile similar but more scaly grey-brown above, with peachy breast and face. Shows thin white wing-bar and square white rump patch in flight. Legs dark, bill dark, longish, downcurved. Call a brisk harsh trill.

Where to see Seashores and shallows in lagoons, lakes and marshland. A widespread passage migrant, most numerous in late summer and autumn.

Temminck's Stint *Calidris temminckii* 14cm

Very small, rather plain sandpiper with short legs, horizontal posture. Grey-brown above, whitish below; recalls miniature Common Sandpiper. Juvenile browner but with very plain appearance; lacks rufous tones and white shoulder-stripes of Little Stint. Long rear end, shortish straight bill, yellow or pale yellow-grey legs. Narrow white wing-bar and broad white tail-sides show in flight. Gives loud trilling flight-call when flushed. Feeds rather slowly with creeping action; unobtrusive. Usually seen singly.

Where to see Freshwater shores. Passage migrant and winter visitor.

Sanderling *Calidris alba* 19cm

Compact, fast-running small sandpiper with short, straight bill, feet notable for lacking hind toe. Breeding birds have plain brownish or russet head and breast, spangled silver and brown upperparts and white belly. In winter becomes very pale silvery-grey on upperparts. In flight shows blackish forewing and black-edged white wing-bar. Bill and legs black. Flight call sharp, fluting *plip*. Dashes back and forth along the shoreline.

Where to see Forages in small flocks on sandy shores. Passage migrant and winter visitor.

Dunlin *Calidris alpina* 19cm

Small sandpiper with short legs and fairly long, slightly downcurved bill. In breeding plumage has black belly patch and rufous mottling on back; upperparts otherwise streaked and speckled grey-brown, flanks white. Bill and legs dark. In winter and juvenile plumage paler, grey-brown with white belly (some dark speckles in juvenile). In flight shows narrow pale wing-bar. Flight call a harsh purring note. Often forms large flocks.

Where to see Muddy seashores and coastal lagoons. Widespread passage migrant and locally abundant winter visitor.

Little Stint *Calidris minuta* 15cm

Very small, short-billed. In breeding plumage pale grey-brown or rufous-tinged above, white below, becoming paler and greyer in winter. Juvenile plumage (most likely in Italy) boldly patterned with dark eye-stripe, pale supercilium and prominent white shoulder-stripe; breast and belly white. Bill and legs dark. In flight shows narrow white wing-bar, broad white tail-sides. Call a short, sharp single note.

juv.

Where to see Muddy shores around coastal lagoons and inland lakes, also on seashore and edges of small pools. Passage migrant and winter visitor.

br.

Jack Snipe *Lymnocryptes minimus* 18cm

Considerably smaller and shorter-billed than Common Snipe, with different head pattern – no central crown-stripe but has short eyebrow-like dark stripe within broad pale supercilium. Plumage otherwise similar to Common Snipe but has more contrasting pattern, with blacker back and more conspicuous yellowish shoulder-stripes. Shy, when flushed usually does not call but drops quickly back down. When foraging, has jerky bobbing action. May associate with Common Snipe.

Where to see Lush marshy fields and drier grassland, lake shores. Scarce passage and winter visitor.

Woodcock *Scolopax rusticola* 35cm

Stout, long-billed, short-legged, short-necked and broad-winged wader. Plumage barred and dappled in shades of rich brown and grey, giving superb camouflage. Distinctive head with eyes set very far back and high up, long sloped forehead and peaked crown with transverse black bars, dark lores and cheek-stripe. Shows reddish rump in flight. Quiet except for territorial male's 'roding' flight at dusk, when he gives sporadic squeaking and croaking calls.

Where to see Breeds in damp woodland with dense understorey; on migration may visit other habitats and even turn up in towns. Mainly winter visitor or passage migrant in Italy; resident in far north.

Great Snipe *Gallinago media* 28cm

Stocky, long-billed wading bird. A little larger than Common Snipe, with proportionately slightly shorter bill. Otherwise very similar, most readily told by different tail pattern, with broad white sides (all brown with narrow white tip in Common Snipe). Also has more heavily barred belly and, when flushed, flies in a straight path (usually silently) rather than zig zagging as Common Snipe does. Shy and skulking, feeds unobtrusively in long vegetation or at water's edge.

Where to see Marshlands, damp pasture. Scarce passage migrant.

Common Snipe *Gallinago gallinago* 25cm

Compact, plump with shortish legs and very long, straight bill. Plumage intricately patterned, face shows dark eye-stripe, pale supercilium, dark crown with pale central stripe. Shows white belly and edges to secondaries in flight. When flushed gives rasping *scaap*. Displaying male in flight produces bleating sound ('drumming') as air vibrates tail feathers. Probes soft ground with long bill.

Where to see Found in marshy habitats (upland and lowland). Mainly a common passage migrant and winter visitor in Italy, breeding in far north.

Red-necked Phalarope *Phalaropus lobatus* 18cm

br.

juv.

Small, delicately built sandpiper, usually seen swimming in buoyant posture. Has colourful breeding plumage; birds in Italy most likely to be juveniles, which have streaky grey and yellow-brown upperparts, whitish underparts, and dark crown and eye-mask. In flight shows narrow white wing-bar and white sides to rump and tail. Black bill straight and very slender, eyes and legs dark. Spins on water and snaps at prey items on surface.

Where to see Sheltered bays, coastal wetlands and inland lakes. Rare passage migrant.

Common Sandpiper *Actitis hypoleucos* 19cm

juv.

Relatively long-tailed sandpiper with horizontal posture and bobbing action. Sandy grey-brown upperparts and mostly white underparts, white curving around bend of wing. Legs greenish, bill fairly short, straight, brownish. In flight shows white wing-bar. Flies with fast, shallow wingbeats interrupted by glides on down-bowed wings. Call a penetrating short whistled *tsweee*, repeated in agitated series in territorial flight low over water.

Where to see Rivers and lakes with gravelly shores, not usually coastal. Widespread but scarce summer visitor, common passage migrant and winter visitor.

Green Sandpiper *Tringa ochropus* 22cm

Larger, taller and darker than Common Sandpiper, with shorter tail and more upright stance; darker than Wood Sandpiper. Upperparts dusky grey-brown with fine white speckles, lower breast and belly white. White eye-rings. Bill longish, dark, legs dull grey-green. In flight shows dark wings, broad-barred tail and white rump patch. Flight call a clear, whistled *tlueet-wit-wit*. Usually feeds alone, moving steadily along shorelines.

Where to see Visits lakes and ponds. A widespread passage migrant and winter visitor, some birds staying on through summer.

Spotted Redshank *Tringa erythropus* 31cm

br.

non-br.

Elegant, slim wader with long legs, long bill with slightly drooping tip. In breeding plumage sooty-blackish with fine white speckles. Winter plumage pale grey with white belly and dark lores, white supercilium. Juvenile similar but darker. Bill dark with red base, legs red. In flight shows long white patch on back. Call a sharp, two-note *che-wit*. Wades up to belly-deep, immersing head or catching flies on the surface.

Where to see Mainly wetlands close to coast. Widespread passage migrant, a few wintering.

Greenshank *Tringa nebularia* 32cm

Large, elegant, pale. Head rather large. Plumage grey on upperparts (darkest on wings), breast finely streaked, belly white. Head pale with prominent dark eyes, bill dark, long and quite stout with slight upturn, long legs greenish. In flight shows dark wings and long white wedge on rump and back, barred tail.

Call a loud *pew*, usually given in series of three notes *pew pew pew*. Wades, occasionally swimming or chasing prey in shallows.

Where to see Freshwater and coastal wetlands; widespread passage migrant, a few winter.

Marsh Sandpiper *Tringa stagnatilis* 23cm

Very elegant and dainty sandpiper with fine, straight bill. In winter plumage, resembles a small version of Greenshank, with light silver-grey upperparts, finely streaked darker, and white belly. Wingtips darker grey, streaked crown. Juvenile similar but browner and more strongly streaked. Legs grey-green, bill and eyes dark. In flight shows all-dark wings, white wedge on rump and back, and dark-barred tail. Call a single whistled *kiu*.

Where to see Wetland shores. Scarce passage migrant.

Wood Sandpiper *Tringa glareola* 20cm

Recalls slimmer, paler version of Green Sandpiper. Brownish on upperparts with pale spangling, this being especially pronounced in juveniles. Breast streaked brown, shading gradually into white belly. Dark eye-stripe and pale supercilium. Legs yellowish, slim straight bill grey, darker at tip. In flight shows all-dark wings and square white rump patch; narrowly barred tail. Call a high three-note whistle, *chiff-if-if*. Forages at water's edge or wades.

Where to see Freshwater and saline marshland with pools. Very common and widespread passage migrant; rarely overwinters.

Common Redshank *Tringa totanus* 26cm

non-br.

br.

Dusky grey-brown on upperparts, fading to paler on belly, underparts spotty in breeding plumage but more uniform in winter. Medium-long slim bill, red with darker tip. Legs longish, bright red (orange in juvenile). In flight shows broad white trailing edge to wing, and white wedge on back/rump. Call ringing, melancholic *teu* or *teu-teu*. Probes for food, flycatches in shallows.

Where to see Lake and river shores, muddy seashores. Nests on wet grassland. Localised breeder, mainly in north-east; widespread passage migrant and winter visitor.

Collared Pratincole *Glareola pratincola* 26cm

Pratincoles are unusual waders, with plover-like plumage and tern-like body proportions – long wings and long forked tail. Collared Pratincole has grey-brown upperparts and breast, yellow-brown lower breast shading to white belly, and yellow throat patch outlined in black. Juvenile has dark scaling on upperparts. Bill is short, stout and slightly downcurved, black at tip, red at base. Eyes large, dark. Has graceful, agile flight, showing white rump, narrow white trailing edge to wing, and reddish-brown underwings. Call sharp, tern-like. Hunts insect prey in air.

Where to see Breeds colonially on flat open ground near water. A localised summer visitor and widespread passage migrant.

Pomarine Skua *Stercorarius pomarinus* 46cm (+10cm tail)

juv.

A large, imposing, barrel-chested seabird. Adult pale morph has dark upperparts including dusky breast-band, pale breast and lower face; dark morph is all dark. Central tail feathers elongated, like twisted spoons. Shows white flash in primary bases in all plumages. Juvenile has short tail and dark grey-brown, mottled plumage, resembling dark juvenile gull. Bill has pale base and dark tip. Powerful flier, may chase and harry other seabirds.

Where to see Offshore, especially near headlands in onshore winds, occasionally on beaches. Uncommon passage migrant and winter visitor.

Arctic Skua *Stercorarius parasiticus* 44cm (+8cm tail)

pale morph

dark morph

Smaller and more elegant than Pomarine Skua, with long falcon-like wings. Adult occurs in pale, intermediate and dark morphs, the former with pale underside, dark cap and dark upperparts, the latter two darker. Shows pale wing flash in all plumages. Central tail feathers elongated and pointed. Juvenile brown with darker mottling, short-tailed, dark-billed. Fast and agile flier, chases other seabirds, especially terns.

Where to see Usually offshore, but also regularly observed closer inshore when parasitising terns; occasionally on inland lakes. Mainly passage migrant and winter visitor, but it can be seen in every season.

Razorbill *Alca torda* 40cm

imm.

imm.

Stocky black-and-white auk with large, vertically flattened bill. Upperparts black with thin wing-bar, breast and belly white; in non-breeding and juvenile plumage lower part of face also white. Adult has vertical white stripe on bill. Usually seen swimming, buoyant, holds short tail slightly cocked. Dives with small jump, wings partly opened. Flies low with rapid wing beats.

Where to see At sea, sheltering in estuary mouths and harbours in severe weather; only likely on beaches if unwell. An uncommon passage migrant and winter visitor.

Kittiwake *Rissa tridactyla* 39cm

A medium-sized, short-legged and long-winged gull. Adult has grey back and wings, small solid black wingtips; in winter develops dark marking behind eye and faint grey 'shawl'. Bill yellow, legs black, eyes dark with narrow red eye-ring. First-winter birds have black head spot, collar and tail-tip, broad black zigzag marking across wings, and black bill. A strong and graceful flyer, more comfortable at sea than most other gulls.

imm. non-br.

Where to see Offshore, sometimes on beaches. Winter and passage visitor but not often seen because of pelagic habits.

Slender-billed Gull *Chroicocephalus genei* 40cm

A slim, pale, medium-sized gull with distinctive tapering head shape and long bill. Back and wings light grey, head and underparts white with pink tint in breeding season. Eyes pale, legs red, bill dark red. Juvenile and subadult have paler bare parts, dark tail-tip and broad light brown wing-bar; never as boldly marked as same-aged Black-headed Gull. In flight wings show broad white leading edge and thin black trailing edge. Call harsh, low *kreeer*. Usually scarcer than Black-headed Gull.

Where to see Breeds in colonies on marshland and islands on coastal lagoons. Localised breeding bird in north-east, Apulia and Sardinia; more widespread on migration and in winter.

Black-headed Gull *Chroicocephalus ridibundus* 37cm

In flight wing always shows broad white leading edge, thin black trailing edge. In summer has brown hood, dark bill and eyes (with thin white eye-ring). In winter bill redder, head white with dark spot behind eyes. Legs red (orange in subadults). Juvenile mottled warm ginger-brown on upperparts. First-winter shows dark wing-bar and tail-tip. Call harsh downslurred screech.

Where to see Breeds on marshland, lake shores and islands, otherwise in all wetland and coastal habitats. Localised breeder, abundant passage migrant and winter visitor.

Little Gull *Hydrocoleus minutus* 26cm

non-br.

Smallest gull, compact with short bill and legs, rather rounded wings. Adult very pale (but shows dark, finely white-edged underside of wings in flight), when breeding has dark hood, otherwise head white with dark crown and spot behind eyes. No white eye-ring, bill and eyes dark, legs pink. Juvenile has dark grey-brown zigzag across wings, dark cap, mask, collar and tail-tip. Has light, tern-like flight, picks prey from water's surface.

Where to see Offshore, sometimes coastal wetlands. Fairly common passage migrant and scarce winter visitor.

Mediterranean Gull *Ichthyaetus melanocephalus* 39cm

Stronger-billed than Black-headed. Breeding adult white with pale grey back and wings, white primaries, jet-black hood reaching nape, prominent white eye-rings. Bill red with black ring, legs red, eyes dark. In winter head white with dusky eye-mask. Juvenile has black legs and bill, grey-brown upperparts. First-winter has dark tail-tip and wing-bar, second-winter like adult but with slight black in primaries. Call an interrogative *kee-yow*.

Where to see Breeds in similar habitats to Black-headed. Summer visitor in north-east; found around all coasts in winter.

Common Gull *Larus canus* 41cm

Long-winged, smaller and darker than Yellow-legged Gull, with dark eyes giving gentler expression. Upperparts mid-grey, with broad black wingtips marked with bold white spots (distinguishing it from Kittiwake); in flight, wings show broad white trailing edge. Bill yellow, legs yellow-green. In winter has streaked head, duller bill and legs. First-winter mottled but with solid grey back. Has strong, light flight. Gives various ringing, nasal yelping and squawking calls.

Where to see Seashores and offshore, coastal wetlands but also inland lakes. Passage migrant and winter visitor.

Audouin's Gull *Ichthyaetus audouinii* 48cm

Rather dusky-looking medium-sized gull. Adult has white head and underparts with pale grey wash, darker grey upperparts, black wingtips with white primary spots. Bill dark red, legs greenish-grey, eyes dark. Juvenile dark, scaly, sooty-grey all over, bill and legs greyish; develops whiter head and some paler grey on back by first-winter. In flight, wings show only a very narrow white trailing edge. Gives various harsh, low-pitched nasal calls. Mainly hunts fish in shallows rather than scavenging.

Where to see Nests in colonies on rocky islands. Uncommon resident on suitable islands around Italian coast, more widespread (especially in south) outside breeding season.

Herring Gull *Larus argentatus* 62cm

A large gull, very like Yellow-legged in all plumages. Adult has silver-grey upperparts, black primaries with white spots at tips, eyes pale with yellow eye-ring (eye-ring orange-red in Yellow-legged), yellow bill with red spot near tip, legs pink. Head becomes streaked in winter. First-winter mottled warm grey-brown, head darker than in same-aged Yellow-legged, bill black. Typical large gull in behaviour, calls include cackling and drawn-out mewing and yelping notes.

Where to see Coasts and offshore, sometimes inland at gull roosts. Uncommon winter visitor.

Yellow-legged Gull *Larus michahellis* 62cm

Commonest large gull. Stout and robust with strong bill, relatively short wings. Adult has white head (lightly streaked grey in winter) and underparts, mid-grey back and wings, black wingtips with white spots. Bill yellow with red spot, legs yellow. Juvenile streaked and mottled grey-brown with pale head, dark bill, pink legs. First-winter similar but paler. Reaches adult plumage in four years. Calls include various low-pitched mews, cackles and grunts.

Where to see Breeds around rocky shores. Widespread in winter, including around towns.

Caspian Gull *Larus cacchinans* 62cm

Resembles Yellow-legged Gull but has subtle structural differences – slimmer and taller, with small eyes, flatter head, longer wings, rather 'saggy' underparts behind legs, bill looks large, long and parallel-edged. Legs duller pinkish-yellow. First-winter has very white head. Has distinctive nasal laughing tone to its various calls. Often dominates other large gull species when foraging in flocks.

Where to see Coasts and inland gatherings of gulls. Scarce but increasing winter visitor, especially in Adriatic Sea and Sicily, expanding its range from eastern Europe.

Lesser Black-backed Gull *Larus fuscus* 62cm

Darker than Yellow-legged Gull in all plumages, smaller and slighter with longer wings. Adult has white head (streaked grey in winter) and underparts, dark grey back and wings, black wingtips with white spots. Bill yellow with red spot, legs yellow (duller in winter). Juvenile dark, rather uniform sooty-grey, black bill, dull pink legs, shows whitish rump in flight. Develops adult plumage over four years. Calls varied yelping, cackling and growling notes.

Where to see Shorelines, fields, rubbish dumps. Fairly uncommon passage migrant and winter visitor.

Little Tern *Sternula albifrons* 23cm

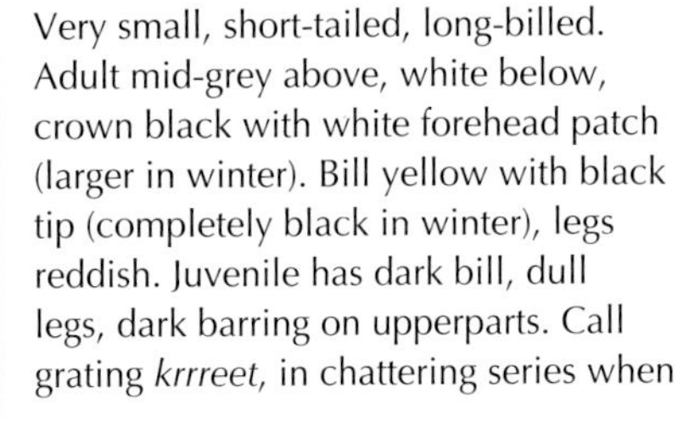

Very small, short-tailed, long-billed. Adult mid-grey above, white below, crown black with white forehead patch (larger in winter). Bill yellow with black tip (completely black in winter), legs reddish. Juvenile has dark bill, dull legs, dark barring on upperparts. Call grating *krrreet*, in chattering series when alarmed. Erratic, jerky flight, picks food from surface.

Where to see Nests colonially on sandy or gravelly seashores, coastal rivers and lakes. Localised summer visitor, mainly in north-east and far south; migrants may be seen offshore anywhere.

Caspian Tern *Hydroprogne caspia* 50cm

A very large, sturdy tern with powerful bill and short tail with shallow fork. Adult in breeding plumage has mid-grey upperparts with darker primary feathers, black cap with slight shaggy crest at rear, and red bill with black marking near tip; legs black. Juvenile and winter adult have white speckles in cap, paler bill. Has distinctive loud, rasping four-syllable call, *kaka-racha*. Feeds by plunge-diving.

Where to see Most likely to be seen at coast but may visit inland wetlands. Uncommon passage migrant, rare in winter. Has bred in Italy once.

Gull-billed Tern *Gelochelidon nilotica* 39cm

Robust, short-tailed, large tern, similar to rarer Sandwich Tern but darker with shorter and much stouter bill, and lacks crest. Adult white on underparts, mid-grey upperparts with darker wingtips, black crown extending down to hind-neck, bill and legs black. In winter head white with restricted dark mask around eye. Juvenile pale with no black cap but sandy tints on wings, neck and crown. Call disyllabic, low-pitched *gur-wick*. Graceful and relaxed flight, hawks insects and picks prey from water's surface but rarely plunge-dives.

Where to see Nests in colonies on open flat ground near water, at coasts and inland. Very localised summer visitor and quite widespread passage migrant.

Black Tern *Chlidonias niger* 24cm

Graceful marsh tern. Breeding adult black with dark grey back and wings. Bill black, legs blackish-red. Underwing pale grey, contrasting with black belly. In winter has white underparts and mostly white head with dark hind-crown and cheek, legs redder. Juvenile similar but with scaly brownish back, pinkish legs. Chattering flight call recalls Little Tern.

Where to see Breeds in marshland, and on migration visits open fresh water; also sea coasts. Scarce, localised and declining breeder in north-west but common and widespread passage migrant.

White-winged Black Tern *Chlidonias leucopterus* 22cm

The smallest marsh tern. Dainty and small-billed. Adult black, paler grey on back, with contrasting white forewings. In flight shows white rump, black underwing. Bill black, legs dark red. In winter white head and underparts, light grey upperparts, dark spot behind eye and grey streaks on crown. Juvenile has dark back, cheeks and hind-crown, white underparts, grey wings. Call soft low *kek* or harsher *chree*.

br.

Where to see Inland wetlands. Breeds irregularly in far north-west of Italy but occurs as regular passage migrant.

Whiskered Tern *Chlidonias hybrida* 26cm

Largest marsh tern. Breeding adult dark grey with black crown and white face. Bill and legs dark red. Winter adult has whitish underparts and white forehead patch. In flight shows silvery uppersides to flight feathers. Tail short with shallow fork. Juvenile has white underparts, brownish scaling on back. Dips to water for prey. Call a shrill rasping note.

br.

Where to see Nests colonially in marshland and river and lake shores. Rare, localised summer visitor to north-east Italy; visits any suitable habitat on migration.

Common Tern *Sterna hirundo* 35cm

Medium-sized, long-tailed tern. Adult has light grey upperparts with darker wingtips, white head with black cap extending to nape, whitish-grey breast and belly. Bill red with black tip, legs short, red. In winter bill darker, forehead white. Juvenile has ginger-brown scaling on upperparts, orange-pink legs and bill base. Call a sharp *kik*; also series of calls *kirri-kirri-kirri*. Dips and plunge-dives for food.

br.

Where to see Nests colonially on gravelly shores or islands, coastal and inland. Summer visitor in north and Sardinia, elsewhere a passage migrant.

Sandwich Tern *Thalasseus sandvicensis* 40cm

Large with long wings and shortish tail. Adult pearly grey on upperwings, with white head, breast and belly. Cap black, shaggy at back. Bill long, black with yellow tip. Legs black. In winter has white forehead. Juvenile has dark barring on upperparts, shorter bill and duskier crown than adult. Call sharp upslurred *kerrr-ick*. Strong flier, performs spectacular plunge-dives.

imm.

Where to see Nests in colonies on sandy beaches or islands, usually coastal. Summer visitor in north-east, otherwise widespread passage migrant and winter visitor.

Red-throated Diver *Gavia stellata* 61cm

Small diver, with slim slightly up-tilted bill. Has blue-grey head and neck and red throat patch in breeding plumage (uncommon in Italy). Bill dark, eyes red. In winter plumage mainly grey above, white below, white neck-sides when seen from behind. Crown grey, eyes prominent on white cheeks. Upperparts white-speckled. In flight looks elongated, feet projecting well beyond body, neck sags slightly. Rarely visits land in winter. Makes lengthy, frequent dives.

Where to see Winter visitor to sheltered northern seas.

Black-throated Diver *Gavia arctica* 69cm

Larger than Red-throated Diver, with heavier, straighter bill. Breeding plumage (uncommon in Italy) dark grey above with large white spots on back, grey head and neck, black throat patch. Winter plumage dull grey above, white below, shows no white neck-sides from behind, grey on crown extends further down cheek, making eyes less prominent. Flight heavier than Red-throated, projecting feet and drooping neck even more noticeable.

Where to see Winter visitor to sheltered bays and estuaries around most coastlines, also further offshore.

Scopoli's Shearwater *Calonectris diomedea* 48cm

Fairly large seabird with large head, very long wings, larger and paler than Yelkouan Shearwater. Upperparts pale grey-brown with lighter feather fringes on back giving scaly impression. Flight feathers darker. Throat and belly white, underwings white with dark edging. Tail short, wedge-shaped, with narrow white rump patch. Bill stout, yellow with dark band. Flies gracefully with glides and relaxed flapping, skimming waves. Also swims, looking buoyant and gull-like on water. Often forages and rests in small groups.

Where to see Nests in colonies on rocky islands throughout Mediterranean. Vocal at nest – loud cawing notes. Most depart in winter but may be seen offshore at any time.

Yelkouan Shearwater *Puffinus yelkouan* 33cm

(Vulnerable) Elegant slim-winged shearwater, much smaller than Scopoli's, with strongly contrasting plumage. Upperparts dark brownish-black, underparts white, with dark wing edges and sometimes dusky sides to undertail. May show dusky half-stripe on base of underwing. Bill long, slim with hooked tip. Flight agile and dynamic, alternating long glides on stiff wings with rapid flapping, tilting to show dark upperside and white underside alternately. Also rests on sea surface. In breeding season comes ashore at night, when very vocal with loud, gargling or coughing calls.

Where to see Nests in colonies in burrows, on islands throughout Mediterranean. Ranges widely throughout the Mediterranean in winter, when may be seen offshore.

European Storm-petrel *Hydrobates pelagicus* 15cm

Very small, delicate-looking seabird. Round-winged and small-billed, with fairly short, square-ended tail – could be mistaken for House Martin at a glance. Blackish-brown with neat white rump patch, fainter whitish band on underside of wing. Has long legs and often patters feet on water's surface, in agile, jerky up-and-down foraging flight. May follow ships. Nests in rock crevices, coming ashore at night, makes purring calls at colonies.

Where to see Breeds sparsely on islets around Sicily and Sardinia; may be seen offshore on migration.

Black Stork *Ciconia nigra* 98cm

imm.

Stately, long-legged bird with long, powerful bill, slightly smaller than White Stork. Black apart from belly, undertail and armpit area, which are white. Has red legs, bill and eye-rings. Juvenile patterned like adult but dark grey rather than black, bare parts duller. Walks in pursuit of prey, in flight glides and soars on long, broad wings. Generally shyer and more flighty than White Stork. Builds stick nest in tree.

Where to see Breeds in quiet, mature forest. Rare, sparsely distributed summer visitor, mainly in south Italy.

White Stork *Ciconia ciconia* 104cm

Unmistakeable, robust large bird with long legs and neck, more stoutly built than herons or Spoonbill. Plumage white, with black flight feathers obvious both at rest and in flight. Legs and long heavy bill bright red (bill dark-tipped in juvenile), eye dark with narrow, short black eye-stripe. Flies with legs and neck outstretched. Quiet, but when reunited at nest pairs perform bill-clattering display. Forages on open ground, stalking prey on foot. Builds very large stick nest; nests reused year on year, and small birds such as sparrows may nest within their structure.

Where to see Often nests in towns or villages on prominent buildings. A localised summer visitor to Italy.

Northern Gannet *Morus bassanus* 93cm

juv.

Large, long-winged, cigar-shaped seabird. Adult white with black wingtips, yellow flush to head, bill and eye-rings grey-blue, feet black. Juvenile entirely dark brown with fine white speckles, gradually acquiring white plumage (first on body, then wings) over four years. Gives harsh grating calls at the nest, otherwise silent. Feeds on fish caught after spectacular vertical plunge-dive; will swim on surface.

Where to see Seen offshore year-round, most commonly in winter. Has bred in Liguria region.

Pygmy Cormorant *Microcarbo pygmaeus* 50cm

Much smaller than Cormorant or Shag, with small, short bill, long pointed tail. Breeding adult blackish with fine white streaks on body plumage. In winter becomes plainer and browner, especially on head, with whitish throat; juvenile paler and browner still, with whitish belly. Nests colonially in trees. Swims low in water, dives for fish, perches with wings spread after fishing trips.

Where to see Wetlands and rivers, does not occur offshore. Increasing and spreading breeding resident, most common in north-east.

Cormorant *Phalacrocorax carbo* 84cm

imm.

Larger, more robust than Shag with thicker bill. Adult blackish with bluish gloss (browner wings), and white chin. In breeding season shows white flank patch and fine white filoplumes on head and neck. Juvenile dull grey, whitish below. Low profile when swimming; slight jump before diving. Gives low croaking calls around nest. Swallows fish at surface. Rests on land with wings spread to dry. Colonial nester.

Where to see Breeds on rocky coasts and around lakes. Localised breeding resident; widespread in winter.

Shag *Phalacrocorax aristotelis* 72cm

imm.

Slimmer than Cormorant with finer bill, steeper forehead. Blackish, glossed green, scaly appearance to upperparts. Develops forehead crest at start of breeding season. Eyes green, bill and legs dark, small patch of yellow skin around gape. Juvenile paler, duller, with whitish chin and belly. Gives low grunts at nest. Swims low in water, jumps to dive, showing long tail. Often rests with wings spread.

Where to see Sheltered rocky bays and further offshore. Fairly common resident in Sardinia, a few pairs breed in the Tuscanian archipelago.

Bittern *Botaurus stellaris* 75cm

Squat, stocky and skulking brown heron. Long dark streaks on throat and breast provide camouflage against reedbed backdrop. Legs relatively short and thick, greenish, toes long. Eyes pale with dark streak below, bill shortish, dull yellow. Stalks slowly along reedbed edges. Neck retracted in flight, broad brown wings giving owlish impression, belied by projecting toes. Croaks hoarsely in flight, male's 'song' a hollow boom.

Where to see Breeds in marshy wetlands with extensive reedbeds; winter wanderers may use smaller reedbeds. Localised resident.

Little Bittern *Ixobrychus minutus* 35cm

♂

♀

Body size similar to Moorhen, with proportionately large head. Adult male light yellow-buff with black crown, back and flight feathers; contrasting pale wing-coverts forming prominent circular patch in flight. Female drabber, juvenile uniform streaky brown. Bill, legs and eyes yellowish. Shy, climbs among reeds with great agility. Song a series of low croaks; gives *kwek* call in flight. Builds nest in thick cover, often above ground level.

Where to see Marshes with reedbeds, wet meadows with scrub. Common but declining summer visitor.

Grey Heron *Ardea cinerea* 90cm

Tall, slender. Grey with whiter underparts, black throat streaks, white crown, broad black supercilium extending into dangling black plumes. Bill and legs yellowish (brighter when breeding), eyes pale. Juvenile has dark grey crown, duller bill and legs. Flies with retracted neck, broad-winged and ponderous in flight. Harsh croaking call. Fairly shy. Hunts at water's edge, waiting for prey or stalking, also in fields.

Where to see Nests colonially in treetops, forages in wetlands. Resident in northern Italy and Sicily, wandering more widely in winter.

Purple Heron *Ardea purpurea* 80cm

Smaller, darker and slimmer than Grey Heron, with strikingly long, thin, snake-like neck and long slim bill. Adult dark blue-grey with reddish-brown tints to forewings, thighs, neck-sides and face. Front of neck and breast marked with heavy dark streaks. Bill, eyes and legs yellowish. In flight the retracted neck forms prominent, angular bulge, toes often spread. Flight call a dry, brief croak. Shyer than Grey Heron.

Where to see Marshes, reedbeds, wet meadows with ditches. Summer visitor and widespread passage migrant.

Great White Egret *Ardea alba* 92cm

Very tall white heron, close to Grey Heron in size but much slimmer build with very long, slender neck – outline helps distinguish from occasional leucistic Grey Heron. In breeding plumage develops long, fringed white plumes or 'aigrettes' on back and on lower breast. Bill yellow (darker in breeding season, with bluish base), feet and legs blackish (developing red tint on upper tibia when breeding). In flight shows very prominent neck bulge, long legs and feet stretching well beyond tail. Rarely calls (except at breeding colonies).

Where to see Visits marshy wetlands and reedbeds. Scarce and localised breeding bird, but increasing; widespread on migration and in winter.

Little Egret *Egretta garzetta* 60cm

Smallish, slender white heron with long, slim neck, often rests in hunched, 'neckless' posture. When breeding, develops long fine plumes on back and breast. Bill dark, with pinkish base when breeding. Legs black, feet bright yellow. Eyes pale. Flight call a harsh squawk. Hunts using curious 'leg-jiggling' action to disturb prey.

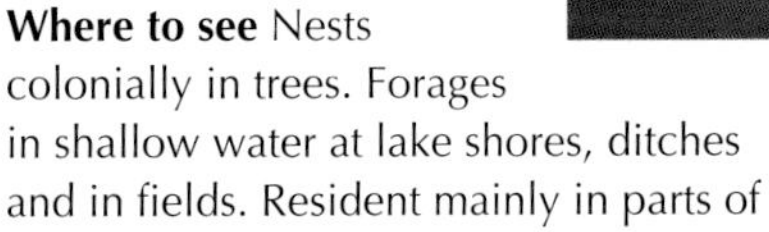

Where to see Nests colonially in trees. Forages in shallow water at lake shores, ditches and in fields. Resident mainly in parts of northern Italy and Sicily, passage migrant or winter visitor elsewhere.

Cattle Egret *Bubulcus ibis* 48cm

non-br.

br.

Small, stocky egret with short bill and full-looking chin. White, with adults developing bright yellowish patches on crown, breast and back in breeding season. Bill yellow (pinkish when breeding), eyes pale, legs pale yellow-grey. Juvenile has dark bill with yellowish base. Flight call a soft croak. Gregarious. Preys on insects disturbed by grazing livestock.

Where to see Fields with livestock; also wet marshland. Nests colonially in trees. Patchily distributed but locally common breeding bird in Italy, wandering a little more widely in winter.

Squacco Heron *Ardeola ralloides* 45cm

Small, pale, thick-necked heron. In breeding plumage yellowish-buff with browner streaked head and back, long dark-edged head-plumes. Bill blue-grey with black tip, eyes pale, legs pinkish. In winter, browner, more heavily streaked, bare parts much duller; juvenile similar. Reveals entirely white wings and tail in flight. Has harsh duck-like flight call. Hunts alone at water's edge, waiting to strike at prey.

Where to see Nests in small colonies in bushes or trees, feeds at marshes, well-vegetated lakes, rivers and other wetlands. Localised summer visitor.

br.

Night Heron *Nycticorax nycticorax* 62cm

Smallish, strong-billed heron. Adult grey, paler below than above, with black back and crown, narrow white forehead stripe, and a few long pale plumes on back. Wings grey in flight. Eyes red, bill black, legs yellow (pinker early in breeding season). Juvenile mid-grey-brown on upperside with heavy pale spotting, underside creamy, streaked darker; eyes, legs and bill dull. Croaking flight call. Most active at night.

Where to see Wetlands, nesting colonially in trees or reedbeds. Fairly common summer visitor, scarce and localised in winter.

African Sacred Ibis *Threskiornis aethiopicus* 68cm

Tall, sturdy with long, strong, down-curved black bill and long black legs. Dirty white with black head, neck and rear. Shows black edge to wing in flight; flies with neck and legs outstretched. Juvenile like adult but dark parts paler sooty grey-black with white speckles. Forages on ground. Sometimes utters harsh croak in flight. Gregarious when feeding and breeding.

Where to see Open country, visits rubbish tips. Feral birds have bred in north-west Italy since 1989; increasing and present throughout north.

imm.

Glossy Ibis *Plegadis falcinellus* 60cm

Large, dark, long-legged bird, shape recalling an oversized Curlew with long downcurved bill. Plumage brownish-black with violet and green sheen, narrow white markings around bill. Juvenile duller and browner, with white speckles on head and neck. Eyes dark, bill and legs dull greyish-pink. Walks and probes at ground when feeding, often seen in small parties. Nests colonially in trees and bushes near water.

Where to see Marshes and reedbeds, damp meadows. A localised breeding bird, more frequent in northern areas.

Eurasian Spoonbill *Platalea leucorodia* 86cm

br.

Graceful, long-legged wading bird with unique long, broad, spoon-shaped bill. Adult white, with yellow tint on neck and long shaggy crest in breeding plumage. Legs black, bill blackish becoming yellowish towards tip, dark lores. Juvenile has pinkish bill, white lores giving more 'open' expression, greyish legs, and shows black tips to primaries in flight. Flies with neck and legs extended. Sleeps in more horizontal position than Great Egret. Feeds by wading (up to belly-deep) and swishing bill through water. Gregarious. Generally silent.

Where to see Found in marshes, lagoons and other wetlands. Scarce breeding bird but widespread passage migrant in Italy.

Osprey *Pandion haliaetus* 56cm

Large, long-winged raptor. Dark brown upperparts, white below with brownish breast-band (heavier in females). Head white with broad dark eye-stripe, shaggy crest at rear of crown. Bill and legs grey, eyes yellow. Juvenile has scaly upperparts; orange eyes. In flight underwing shows barred flight feathers, dark wrist patch. Relaxed flapping flight recalls large gull. Gives yelping calls around nest. Hovers and dives feet-first to catch fish.

Where to see Hunts at lakes and sea coasts. Rare summer visitor, a few pairs breeds in Sardinia and Tuscany; scarce but more widespread as passage migrant.

European Honey-buzzard *Pernis apivorus* 55cm

Very variable raptor, lankier and smaller-headed than Common Buzzard. Adult typically light brown with grey head, underparts mottled. Feet and eyes yellow, bill relatively small and long. Tail has broad bars at base and tip with wide gap between. Juvenile very like Common Buzzard; dark-eyed, plumage varies from pale through rufous and grey-brown to very dark. Gives whistling calls near nest. Feeds mainly on wasp larvae.

Where to see Breeds in woodland with open countryside nearby. Very widespread summer visitor and passage migrant.

Egyptian Vulture *Neophron percnopterus* 60cm

The smallest vulture, with wide rectangular wings and longish, diamond-shaped tail. Close views reveal long, shaggy feathers at back of head. Adult mostly white. Flight feathers are solid black on underside, black-bordered on upperside. Bare face may be pale or bright yellow, bill relatively small. Juvenile all dusky brown, acquiring white adult plumage over five years. Gregarious, visits carcasses and rubbish dumps.

Where to see Rugged open countryside. Very rare and local as summer visitor in southern Italy and Sicily.

Bearded Vulture *Gypaetus barbatus* 110cm

A very large, striking bird of prey, with distinctive silhouette due to long, wedge-shaped tail. Adult has dark grey upperparts and warm buff head and underparts, with black eye-mask that extends into 'beard', eyes pale. Juvenile has dark head and dusky whitish underparts. Flies high in slow and graceful manner, usually seen alone or in pairs. Visits carcasses and drops bones (and tortoises) from height to break them open.

Where to see Mountains and high uplands. Resident in the Alps.

Griffon Vulture *Gyps fulvus* 100cm

A huge, scavenging bird of prey with dull sandy-brown plumage, head and long neck downy white, whitish neck-ruff. Eyes dark, bill yellow. Juvenile has grey bill and brownish ruff. In flight shows very long, broad wings with outer primaries separated as long 'fingers'; flight feathers and tail dark brown. Makes hisses and grunting sounds at nest and when gathered at carcasses.

Where to see Prefers rugged countryside with mountainsides to generate thermals. Breeds in Sardinia but increasingly seen throughout Italy because of reintroduction projects, with new breeding colonies established from the Alps south to central and south Italy and Sicily.

Short-toed Eagle *Circaetus gallicus* 65cm

juv.

Stocky, big-headed, long-winged and short-tailed. Fairly uniform brown above and pale with variable barring below; pale and darker forms exist. In flight often looks very pale from below, with darker breast and head. Underwing shows evenly spaced barring, no dark patch at wrist; tail has well-spaced bars. Bill and feet grey, eyes yellowish-orange. Has fluting call when breeding. Often hovers before dropping onto prey.

Where to see Breeds in open, rugged countryside with some trees. Fairly widespread summer visitor and passage migrant.

Greater Spotted Eagle *Clanga clanga* 65cm

A large but compact dark eagle, with broad wings and rather short tail. Adult rather uniform dark brown with slightly paler band on underside of wing. Juvenile has pearly white spots on tips of wing feathers; juveniles also occur in a pale morph with fawn head and body, and darker flight feathers. Eyes dark. Glides with hand of wing slightly downturned, spends much time resting on well-hidden perches.

Where to see Any wild open country, including near water. Rare winter visitor.

Booted Eagle *Hieraaetus pennatus* 46cm

dark morph

light morph

Buzzard-sized with long wings, longish, square-cut tail. Occurs in dark, pale and intermediate morphs, body rather uniform with slight darker streaking on upperparts and underparts. Eyes pale, legs feathered, feet yellow. In flight from below, pale morph shows contrast between pale body and forewing and dark flight feathers. Upperwing shows pale mid-wing patch. Gives shrill and mewing calls near nest. Soars and stoops on prey.

Where to see Open country and light woodland; uncommon passage migrant and winter visitor. A few oversummer in south.

Bonelli's Eagle *Aquila fasciata* 65cm

juv.

Graceful and powerful large eagle with long wings, and long tail giving hawk-like silhouette. Rather pale silver-grey on head and body with darker grey-brown wings, from below shows obvious wide dark wing-band and tail-tip, from above shows silvery back. Flight feathers strongly barred. Eyes yellow, legs feathered. Juvenile variable but usually warm brown, lacks adult's contrasting wing pattern. Often seen gliding in pairs, seeking prey.

Where to see Montane regions, also other rugged and wild countryside including wetlands. Resident on Sicily.

Golden Eagle *Aquila chrysaetos* 86cm

Very large, imposing and majestic raptor, well-proportioned with long wings and relatively long tail, orange eyes. Adult (5+ years old) dark grey-brown with bright reddish-orange streaks on head and neck, some paler yellowish feathers on forewing, forming vague bar in flight. Juvenile more uniform warm dark brown; in flight shows large white wing patch at primary bases; white areas gradually shrink with successive moults. Generally quiet but may give soft whistling call. Soars easily on thermals and skims along crag edges, searching for prey or carrion. Adult pairs hold large territory over many years; subadults wander.

juv.

Where to see Prefers remote mountainous areas with some forest. Widespread resident in uplands.

Marsh Harrier *Circus aeruginosus* 49cm

The largest and stockiest harrier. Long-winged and long-tailed. Male has light grey head, body, tail and most of wings, red-brown back, belly and inner wings, large black primary patch, eyes pale. Female and juvenile very dark brown with variable creamy crown, chin and shoulders, eyes dark. Hunts in low, steady patrolling flight over vegetation, listening and watching for prey moving on ground.

Where to see Marshland, reedbeds, wet pasture, sometimes farmland. Resident in north, passage migrant and winter visitor further south.

Hen Harrier *Circus cyaneus* 50cm

Slimmer than Marsh Harrier, with longer wings and tail. Male light grey with large black wingtips and white rump, eyes yellow. Juvenile and adult female streaky grey-brown with white rump, banded tail, pale outline to facial disc; eyes dark in juvenile, paler in adult female. Very light, buoyant low flight, locating prey on ground mainly by sound. Forms communal roosts in winter and on migration.

Where to see Hunts over well-vegetated open countryside, farmland, meadows, heathland, marshes. A widespread passage migrant and winter visitor.

Pallid Harrier *Circus macrourus* 45cm

Small harrier. Male very pale ghostly grey, with faint white rump, and black wingtips (more restricted than in Hen Harrier). Female has mottled brown upperparts, pale underparts with strong dark streaking. Juvenile has unstreaked warm rufous underside, very like juvenile Montagu's Harrier, with stronger face pattern and darker 'neck shawl'. Searches fields in low 'quartering' flight when hunting, may join communal roosts with other harriers.

Where to see Open countryside, including farmland and wetlands. Uncommon passage migrant and rare winter visitor.

♂

♀

juv.

Montagu's Harrier *Circus pygargus* 45cm

♂

♀

Small harrier. Very long-winged and elegant; shows four 'fingers' on wingtips in flight (five in Hen Harrier). Male grey with extensive black wingtips, narrow white rump patch, faint reddish streaks on lower belly. In flight shows conspicuous single black wing-bar, doubled on underwing. Female has streaked underparts, juvenile rufous and uniform on underparts; both have banded tail and white rump. Gives chattering calls near nest.

Where to see Open countryside, arable farmland, meadows, marshland. Summer visitor, more common in north, and widespread passage migrant.

Goshawk *Accipiter gentilis* 52–61cm

Large, powerful raptor. Adult dark grey-blue above, whitish below with fine darker barring, white supercilium and undertail. Eyes orange. Juvenile brown, pale below with dark streaks; eyes yellow. In flight markedly stockier than Sparrowhawk with large protruding head, round-ended rather than square-cut tail, distinct 'bulge' to trailing edge of secondaries. Has fast low hunting flight; in courtship soars with white undertail puffed out. Hunts birds and mammals.

Where to see Undisturbed woodland near open countryside. Fairly widespread resident, wandering more widely in winter.

Sparrowhawk *Accipiter nisus* 32–38cm

Female much larger than male. Long-tailed and broad-winged. Adult male blue-grey above, brick-red on cheeks and breast, becoming barred red on whitish belly. Adult female grey-brown above; underside pale, finely barred darker. Juvenile brown with pale fringes on upperparts, coarse brown barring on underparts. Eyes yellow (becoming redder in older males). Gives chattering calls near nest. Flight alternates fast flaps and short glides; also soars, especially early in breeding season. Hunts birds.

Where to see Found in woodland, parks, gardens. A widespread resident.

♀

♀

♂

Red Kite *Milvus milvus* 66cm

(Near Threatened) Large but relatively lightweight raptor with long wings, unique long, deeply forked tail. Plumage warm rufous-brown, mottled on upperparts, streaked on underparts. Head greyish with darker streaks. Eyes pale. Underwings dark with black wingtips and pale 'window' at primary bases. Call a weak mewing. Agile in flight, soars and glides, twisting tail to steer. Feeds mainly on carrion; may gather in large numbers.

Where to see Open countryside with some woodland. Widespread resident in central and southern Italy, passage migrant and winter visitor further north.

Black Kite *Milvus migrans* 53cm

Like Red Kite but smaller, with shorter tail and duller, less contrasting plumage. Adult almost uniform grey-brown, eyes pale. Juvenile has dark eyes, mottled upperparts, suggestion of dark eye-mask. In flight, underwing shows vague paler window at primary bases. Tail only slightly forked and looks square-ended when fanned. Gives mewing and whistling calls. Graceful on the wing though a little more laboured than Red Kite. Often seen in large gatherings.

Where to see Visits ploughed fields to take worms, and joins crows and gulls at rubbish dumps; also scavenges around fisheries. Nests in woodland. A common, widespread summer visitor and passage migrant.

White-tailed Eagle *Haliaeetus albicilla* 88cm

Very large, stout eagle with short tail, long, rectangular wings and very large bill. Looks shaggy and unkempt when perched. Adult dark brown with paler head, yellow bill, pale eyes and pure white tail. Juvenile (most likely in Italy) has darker head, bill and eyes, and dark borders to pale tail feathers, gradually attaining adult plumage over at least four years. Mainly a scavenger and sometimes gregarious.

Where to see Coastal lowlands, farmland, other open countryside. Rare winter visitor in central and northern Italy.

Common Buzzard *Buteo buteo* 55cm

juv.

Large, familiar raptor, compact with broad wings and short tail. Plumage basically mottled dull brown but highly variable, from almost white to extremely dark. Most individuals dark brown with more or less prominent paler breast-band. Tail evenly barred, underwing shows dark feather tips and dark patch at wrist-bend (prominent in pale birds). Eyes dark (paler in juvenile). Gives mournful, far-carrying mewing call. Soars easily on thermals, often in groups; also often seen resting on low posts. Very diverse diet, including carrion and live prey (vertebrates and invertebrates).

Where to see Found in open, mainly lowland countryside with some trees. Common and widespread resident.

Long-legged Buzzard *Buteo rufinus* 58cm

Usually larger than Common Buzzard, and (although variable) usually paler and more reddish in plumage. Most adults have very pale head and breast, shading darker on belly, with brown, darker-mottled upperparts. Juvenile generally darker and more uniform than adult. In flight, underwing shows dark tips to flight feathers and dark carpal patch but otherwise pale. Soars in thermals, searching ground below for carrion or prey, or rests on low perches.

Where to see Open, often rugged countryside, farmland. Rare passage migrant.

Eagle Owl *Bubo bubo* 68cm

Very large, sturdy, imposing owl. Nocturnal. Dense plumage warm brown with close-spaced mottling on upperparts, heavy streaking on underparts. Broad head with prominent ear-tufts, eyes orange. In flight looks very broad-winged, shows somewhat pale underwing with dark tips and barring to flight feathers. Territorial song a loud, far-carrying deep, single hoot; also gives yelping calls, and chicks beg with intense wheezing calls. Hunts birds and mammals.

Where to see Remote and rocky habitats with some trees. Resident, quite common in the Alps, uncommon elsewhere.

Barn Owl *Tyto alba* 36cm

A slim, pale owl with long wings and legs. Upperparts mottled golden and grey, underparts white or with buff wash (female typically darker). Long oval or heart-shaped facial disc white, with thin dark border. Eyes black, relatively small. In flight, looks ghostly; underwing pale and almost unmarked. Call a harsh, hoarse scream. Has buoyant, slow hunting flight, turning and hovering often before dropping down with legs outstretched. Active mainly at night but sometimes in daylight. Hunts small mammals, which it locates by sound.

Where to see Prefers open country with some derelict buildings or old trees with holes for nesting. Common resident.

Eurasian Scops Owl *Otus scops* 20cm

Very small, slender owl with ear-tufts. Plumage tone varies from rufous-brown to greyer, but always finely and intricately patterned; superbly camouflaged against tree bark. Eyes yellow. Face and body shape and prominence of ear-tufts changes drastically with bird's state of alertness – at rest becomes tall and slim with tufts fully raised. Male's song a soft, tuneful purring note, repeated constantly. Strictly nocturnal, resting in cover by day. Feeds mainly on insects, hunts by pouncing from a perch.

Where to see Found in all kinds of wooded areas including town parks and gardens; often nests in old woodpecker holes. A widespread, common summer visitor, resident in far south.

Pygmy Owl *Glaucidium passerinum* 17cm

Very small, round-headed owl with pointed tail that it often holds cocked. Upperparts dark grey-brown with white speckles, underparts whitish with thick grey-brown streaks. Tail barred. Has yellow eyes under short, frowning white brows, set in poorly defined facial disc. May be active by day or night. Gives repeated whistling calls, recalling Bullfinch. Nests in old woodpecker nest-holes. Very effective hunter of small birds and mammals.

Where to see Forests with pine trees, also forest edges. Quite common resident in the Alps, uncommon in the western part.

Little Owl *Athene noctua* 25cm

Small, stocky with short wings and tail, broad, flat-topped head. Upperparts grey-brown with white spots (small on crown, large on back and wings. Underparts pale with darker streaking. Has large yellow eyes under frowning white supercilia. Facial disc not well-defined. Juvenile more uniform in plumage tones, with unspotted crown. Has bounding flight pattern. Call a high-pitched two-note *kyew*. Pounces on prey from perch. Most active in early evening.

Where to see Open countryside; farmland, scrub, gardens; often nests in old buildings. Common resident.

Ural Owl *Strix uralensis* 56cm

Large with a round head, relatively long tail, no ear-tufts. Plumage mottled and streaked in dark and light grey. Facial disc light grey, eyes small and black, bill yellow. Male's territorial call a series of low, mellow hoots; female's more rasping. Both sexes also give sharp *kuwick* contact calls. Nocturnal, hunts small to medium-sized birds and mammals. Nests in tree cavities, nestboxes or old nests of other large birds.

Where to see Mixed montane woodlands. Rare but increasing resident in far eastern Alps.

Tawny Owl *Strix aluco* 40cm

Fairly large, big-headed owl. May be greyish, brown or rufous. Large facial disc, black eyes prominent. Young leave nest early, still flightless and fluffy. Very round-winged in flight, shows evenly barred flight feathers and no obvious dark marking at wing-bend. Song a long, fluting, quavering hoot, call a sharp *ke-vick*. Hunts mainly from perch, takes great variety of prey. Sedentary and territorial, uses large tree hole for nesting.

Where to see Woodland. A common resident but absent from far south-east and Sardinia.

Long-eared Owl *Asio otus* 34cm

Slim, long-winged with prominent ear-tufts. Upperparts grey-brown, mottled darker, line of white spots on shoulder. Underparts buff with dark streaks. Facial disc rufous with whitish brows and bill base, eyes orange. Fledglings downy grey with black face. Orange patch at primary bases on upperwing; underwing paler with black carpal marking. Territorial song is repeated short, deep hoots; chicks make 'squeaky-gate' noises. Hunts in flight at night. Usually nests in old birds' nests.

Where to see Forest edges, plantations, and other semi-open habitats. Widespread resident.

Short-eared Owl *Asio flammeus* 38cm

Medium-sized, long-winged owl, often active in daylight. Plumage sandy grey-brown with dense darker mottling on upperparts, finer streaking on underparts. Upperside of wing shows yellowish patch at primary bases, underwing very pale with dark feather tips and small carpal patch. Hunts in flight, hovering before dropping onto prey (mainly voles); several may patrol the same field.

Where to see Open rough grassland, including around wetlands. Uncommon passage migrant and winter visitor (but numbers vary greatly from year to year).

Tengmalm's Owl *Aegolius funereus* 25cm

juv.

Small, big-headed, strictly nocturnal owl. Upperside dark grey with large pearly white spots, underside pale with cross-hatched streaking. Facial disc whitish, boldly outlined black, with raised black markings above bright yellow eyes giving startled expression. Juvenile dark brown with a little white around eyes. Territorial song a fast series of short barking hoots, also gives sharp *kip* call when alarmed. Hunts small mammals and birds.

Where to see Forested habitats, especially coniferous. Quite common resident in the Alps, uncommon in the western part.

Hoopoe *Upupa epops* 27cm

Unmistakeable, medium-sized bird with long crest and long, slender downcurved bill. Head and body plumage soft pinkish-buff, wings and tail boldly patterned in black and white. Crest usually held swept back to form point but can be raised and spread; each feather tipped black. In flight, looks very broad-winged, giving butterfly-like impression. Song a three-note hoot, also various harsh rolling calls. Catches insect prey on ground. Surprisingly inconspicuous when moving slowly in dappled shadow.

Where to see Prefers short grassland for foraging but needs tree hole or crevice in building for its nest. A common and widespread summer visitor.

Kingfisher *Alcedo atthis* 18cm

Compact, very colourful bird. Short-tailed, long-billed and very short-legged. Upperparts blue-green with fine paler spots; rump and back dazzling iridescent light sky-blue. Underparts orange. White stripe on neck-side and white chin, orange patch behind eye, head otherwise blue-green. Bill dark, with red on lower mandible in female. Eyes dark, feet red. Juvenile slightly duller with dark feet, patchy brownish breast-band. Call a high, piercing single whistle. Preys on fish and other aquatic animals which it catches with plunge-dive from perch or after hovering. Nests in bankside tunnel.

Where to see Found around rivers and lakes, sometimes sheltered sea coasts. Common, widespread resident.

European Bee-eater *Merops apiaster* 27cm

Spectacular, unmistakeable multicoloured bird, slim and elongated with large head. Blue underparts and flight feathers, orange-brown upperparts with yellow shoulders, yellow throat outlined in black. Bill long, downcurved, eyes red. Central tail feathers protrude as long spike. Juvenile a little paler and drabber, green-tinged on upperparts. Very graceful in flight; pointed wings are pale underneath with black trailing edge. Call a pleasant rolling *prrrt*. Catches insect prey on the wing. Gregarious, nests in colonies in burrows.

Where to see Open scrubby countryside in warm lowlands with plenty of insect life. Nests in sandbanks or riverbanks. A widespread summer visitor.

juv.

♂
♀

Roller *Coracias garrulus* 31cm

Robust bird with large head, strong bill. Somewhat crow-like in shape, but dazzlingly colourful. Body and most of wings bright iridescent turquoise-blue. Back and inner parts of wing bright chestnut. In flight looks Jackdaw-shaped, flight feathers blackish above, dark blue below. Tail has dark base and black corners. Eyes and bill black. Juvenile is duller, paler version of adult. Calls (including alarm call) are harsh crow-like notes. Hunts insects and small vertebrates, often watching for prey from perch on overhead wire.

Where to see Open countryside with some trees; usually nests in tree-hole. Patchily distributed summer visitor, more widespread on migration.

Wryneck *Jynx torquilla* 17cm

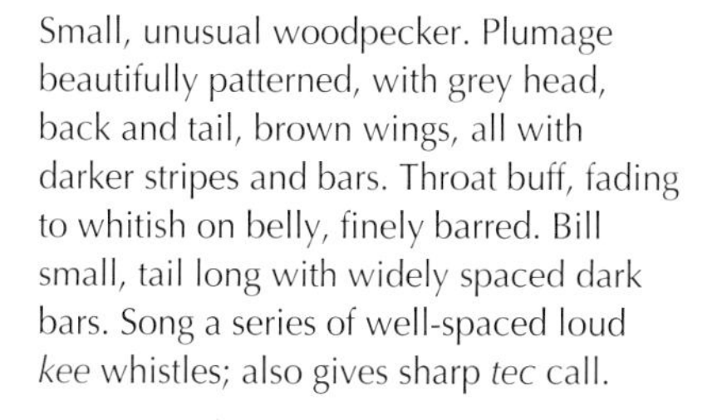

Small, unusual woodpecker. Plumage beautifully patterned, with grey head, back and tail, brown wings, all with darker stripes and bars. Throat buff, fading to whitish on belly, finely barred. Bill small, tail long with widely spaced dark bars. Song a series of well-spaced loud *kee* whistles; also gives sharp *tec* call.

Does not climb tree trunks but forages on ground, searching for ants, hopping like a thrush. Uses existing tree-holes for nest.

Where to see Woodlands, countryside with trees, city parks and large gardens. Widespread summer visitor, scarce in winter.

Three-toed Woodpecker *Picoides tridactylus* 17cm

A smallish, stocky, black-and-white woodpecker. Face black with white chin and double white stripe above and below eyes. Back and underside white with some black barring, wings and tail mostly black. Male has yellow patch on crown, female's crown black with white streaks. Has sharp *kik* call, drums in rather long double bursts. Drills holes in tree-trunk bark to access sap.

Where to see Found in montane coniferous and mixed woodland, more frequent near tree-limit. Resident in north-east Alps.

Middle Spotted Woodpecker *Dendrocoptes medius* 21cm

Whiter than other 'pied' woodpeckers. Crown red, undertail pink. Has black cheeks/throat-side patches not reaching back of neck, and fine dark streaking on otherwise white underparts. Wings black with bold white barring and large white shoulder patch. Bill rather small. Active, spending more time in higher twigs and branches than larger species. Calls like those of Great Spotted but softer; rarely drums.

Where to see Prefers deciduous woodland, especially oak. Localised resident in central and south Italy and near Trieste.

White-backed Woodpecker *Dendrocopos leucotos* 27cm

♀

The largest 'pied' woodpecker. Wings black with broad white barring, back white (but often mostly concealed by folded wings. Underparts white with black streaks and red undertail. Face/throat-sides pattern like Great Spotted but cheek stripes do not reach back of neck. Male has red crown, female's is black. Call a low-pitched *kuck* or quiet purr. Drums in long, accelerating bursts. Shy, somewhat slow-moving.

Where to see Mature deciduous forests with plenty of standing deadwood. Scarce resident in mountains of central Italy, especially Abruzzi region.

Great Spotted Woodpecker *Dendrocopos major* 24cm

Starling-sized black-and-white woodpecker with strong bill. Black upperparts with white bars on wings and long oval patch on shoulders. Undertail red. Head white with black crown and stripes on lower face, enclosing white cheek. Male has red patch at hind-crown. Juvenile has red crown (more extensive in male juveniles) and pink undertail. Call a sharp *kik*, series of longer grating notes in territorial dispute. Drumming usually lasts less than 1 second, beats very fast, blending into single 'creaking' sound. Feeds in trees, breaking into decaying wood to access grubs.

Where to see Woodland, particularly mixed with some spruce and pine. Common, widespread resident, including Sicily and Sardinia.

♂

juv.

Lesser Spotted Woodpecker *Dryobates minor* 15cm

Very small, short-billed and compact black-and-white woodpecker. Black upperparts broadly barred white. Underparts whitish with fine dark streaks; no red undertail. Head white with black stripe from bill base curving around cheek; crown red with black border in male, solid black in female. Call is weaker version of Great Spotted's, also a series of ringing notes, recalling Wryneck. Drumming soft, slower than Great Spotted's. Climbs and forages along thin high twigs.

Where to see Fairly common, widespread woodland resident on mainland.

Grey-headed Woodpecker *Picus canus* 25cm

Distinctly smaller than Green Woodpecker with less colourful face markings. Head lead-grey with black lores and narrow black stripe from bill base; male has red patch at front of crown. Underparts greyish with faint barring on undertail, upperparts dull green with brighter, more yellowish rump. Call a series of weak piping notes. Drums regularly but softly. Feeds on ground and in trees, readily comes to bird feeding stations.

Where to see Deciduous woodland and adjacent more open areas. Resident in eastern Alps.

Black Woodpecker *Dryocopus martius* 43cm

Very large woodpecker; crow-sized but different shape with long slim neck, slight shaggy crest at rear of crown giving angular head shape, pointed tail. Plumage black except for red crown in male, red hind-crown patch in female. Eyes and bill pale. Looks elongated in flight, and flies directly rather than bounding like smaller woodpeckers. Various loud ringing calls, and very loud far-carrying drumming in 2–3-second bursts. Forages mainly on tree branches and trunks, excavating grubs from rotting wood. Nest-hole oval, 12cm across at widest point.

Where to see Found in mature forest. Common in Alps, scarce in Apennines and far south, absent from Sicily and Sardinia.

♂

♀

Green Woodpecker *Picus viridis* 33cm

Large, rangy-looking. Body plumage dull olive-green, paler grey-green on underparts. Crown and hind-neck red. Black eye-patch and broad black stripe below cheek; red-centred in male, black in female. Eyes pale. Juvenile very spotty, lacks black on face. Bill sturdy, horn-grey. In flight shows vivid yellow-gold rump. Call loud series of yelping, laughing notes. Seldom drums; drumming roll quiet, lasts 1.5 seconds. Feeds mainly on ants, hops on ground.

Where to see Open countryside with mature trees for nesting. Common, widespread resident on mainland.

Common Kestrel *Falco tinnunculus* 34cm

Slightly larger than Lesser Kestrel. Male has grey tail with black tip, and grey head with dark, narrow moustachial stripe. Upperparts chestnut with black spotting, underparts light buff with dark spots. Female chestnut on upperparts with strong black barring, underparts pale, streaked darker, tail barred. Call a series of short, sharp *kee* notes. Has fast flapping flight, hovers, sometimes soars. Less gregarious than Lesser Kestrel. Nests in tree hollow or on building.

Where to see Hunts over grassland, farmland, road verges. Common, widespread resident.

Lesser Kestrel *Falco naumanni* 30cm

Small, elegant, long-tailed falcon. Male has solid blue-grey head, wing panel and tail, unmarked chestnut back and forewing, black tail-band and primaries. Chin white, underparts light buff with fine dark spots. Female chestnut above with black spotting, paler below with dark streaks, tail banded. Eyes dark, feet yellow, claws white (black in Common Kestrel). In flight, shows long pointed wings, central tail feathers slightly longer. Has distinctive rasping three-note call. Hunts insects and small vertebrates, hovering before dropping down. Gregarious, nests in loose colonies.

Where to see Nests mainly in old buildings (including in towns), hunting in nearby open countryside. Uncommon but increasing summer visitor, mainly in south.

Red-footed Falcon *Falco vespertinus* 31cm

(Near Threatened) Male is uniform dark slaty-grey except for thighs and undertail, which are dark reddish. Female has dark grey back and wings, with black chequering, white face with dark eye-mask and moustachial stripe, and rich orange-buff crown and underparts. Legs and eye-rings red. Juvenile like paler, browner juvenile Hobby. Call, given around nest, rapid *kekekeke*. Hawks insects in flight, Hobby-like, but also hovers and catches prey on ground.

Where to see Summer visitor to Po Valley; also fairly common and widespread passage migrant.

♀

juv.

♂

Eleonora's Falcon *Falco eleonorae* 39cm

Fairly large, long-tailed and very long-winged falcon. Adult has two colour morphs: dark (uniform brownish-black) and pale (dark upperparts, white cheeks, red-brown underparts with dark streaks). Juvenile is greyer, scalier version of pale adult. Call repeated sharp, grating *keh* calls. Fast flickering flight, very agile; catches small birds on wing over open sea. Begins breeding cycle very late in summer, so chick-rearing coincides with peak passerine migration through region.

Where to see Nests in colonies on rocky islets and sea cliffs. Summer visitor, present well into autumn, most numerous around Sicily and Sardinia.

dark morph

light morph

Merlin *Falco columbarius* 27cm

Small, compact, very fast-flying falcon, proportions similar to much larger Peregrine. Adult male silver-grey on upperparts with fine dark streaking, underparts pale buff with dark streaks. Female and juvenile have dark grey-brown upperparts, pale underparts with heavy, blocky streaking. Call similar to Common Kestrel's. Usually hunts low over ground, pursuing and ambushing small birds; may effect bounding thrush-like flight to deceive prey.

Where to see Most likely in undisturbed, open coastal habitats with scrub and grassland. Uncommon winter visitor.

Lanner Falcon *Falco biarmicus* 47cm

Slightly smaller and slimmer than Peregrine. Upperparts dark grey-brown with paler fringes. Crown grey with rufous patch at hind-crown. Cheeks white with narrow dark moustachial stripe. Undersides white with black spotting, flanks barred. Juvenile browner; underparts washed buff and marked with heavy dark streaks. Flight feathers and tail barred. Gives harsh, rasping calls. Mainly hunts birds. Nests on cliff ledge or in old bird's nest in tree.

Where to see Open, arid countryside. Widespread but uncommon; absent in north and on Sardinia.

Hobby *Falco subbuteo* 32cm

Close to Kestrel in size but with much longer, narrower wings. Adult dark slate-grey above with white chin and cheeks, narrow black moustachial stripe and shorter black 'spur' behind, creating distinctive face pattern. Underparts pale with heavy, neat black streaks; underwing heavily chequered. Undertail and thighs rufous-red. Juvenile similar but browner, upperparts scaly, face more dusky, buff rather than red undertail and thighs. Call comprises ringing *kew kew* notes. Agile, tireless wheeling flight, recalling a swift; catches and eats large insects on the wing. Nests in old bird's nests in tree.

Where to see Hunts over open ground, marshes, heaths. A fairly widespread summer visitor and passage migrant.

Saker Falcon *Falco cherrug* 52cm

A very large, powerfully built falcon with long tail. Head and underparts pale, with darker streaking that becomes dense and heavy on lower belly, flanks and leg 'trousers'. Upperparts dark brown, showing paler feather fringes in fresh plumage, with darker flight feathers and paler, densely barred tail. Underside of wing strongly barred. Mainly takes prey on ground, favouring small mammals. Often chooses high, prominent perches from which to watch for prey.

Where to see Rugged and low-lying open country. Rare visitor, mainly in winter.

Peregrine Falcon *Falco peregrinus* 42cm

imm.

Large, powerful and stocky. Adult dark slate-grey above, white below with black barring. Dark crown and broad dark moustachial stripes stand out against white cheeks. Underwings finely barred. Juvenile browner, underparts buff-tinged with heavy dark vertical streaks. Looks broad-winged in flight; rump pale and silvery-blue. Call a loud, harsh and whining *kreeh kreeh*, very vocal near nest. Stoops on birds from height.

Where to see Nests on cliff ledge, sometimes on tall buildings. Hunts over open countryside, especially coastal. Fairly common widespread resident.

Rose-ringed Parakeet *Psittacula krameri* 40cm

Brilliant green, very long-tailed parrot, native to Asia and Africa. Male has black chin and black collar with pink edge. Eye-rings and bill red, feet grey. In flight, shows darker flight feathers than inner wing. Blue and yellow forms occasionally seen. Call a harsh screech. Feeds on buds, fruit and seeds. Mainly forages high in tree canopy, climbing among branches; manipulates food with foot. Nests in tree-holes. Gregarious outside breeding season.

Where to see Escapees established in some city parks, including in Rome, Milan, Bologna, Genoa, Florence, Palermo. Rapidly increasing.

Monk Parakeet *Myiopsitta monachus* 29cm

Smaller and shorter-tailed than Rose-ringed Parakeet. Plumage vivid mid-green, with a paler and greyer lower face and underside, barred slightly darker. Flight feathers glossy blue-green. Eyes dark, bill yellowish-pink. Calls squeaky chattering and squawking. Eats fruits, seeds, buds and other plant matter. Highly gregarious. Builds large communal stick nests in trees, pairs nesting in individual 'apartments' inside.

Where to see Native to South America. Established from escapees in some city parks, including Genoa and Rome where it is quite common.

Golden Oriole *Oriolus oriolus* 23cm

♀

Colourful but surprisingly hard-to-see bird. Robust shape, recalling miniature crow. Male bright yellow with black tail, wings and lores, yellow patch midway down folded wing, and yellow tail corners. Female drabber, yellow-green with darker wings; grey-white below with darker streaks; juvenile greyer and more uniform. Eyes and bill dark pink, legs grey. Song a loud, exotic-sounding fluty whistle; also has harsh crow-like calls. Forages and nests in high treetops, where it is well camouflaged among foliage.

Where to see Found in mature deciduous woodland, often near watercourses, also in large parks and gardens. A fairly widespread summer visitor.

♂

Red-backed Shrike *Lanius collurio* 17cm

Male has grey cap and black eye-mask, chestnut back and wings, dark tail, white underparts with rosy flush. In flight, shows distinctive tail pattern: broad black terminal band and central stripe, sides white. Female and juvenile like very drab, grey-brown male, with scaly underparts. Calls short, harsh notes, song quiet, warbler-like with mimicry. Hunts large insects, sometimes impaling them on thorns. Watches for prey from prominent perch.

Where to see Open scrubby countryside; farmland, orchards, olive groves, vineyards. Common and widespread summer visitor.

Great Grey Shrike *Lanius excubitor* 23cm

Large, long-tailed. Crown and upperparts grey, wings and tail black. Has narrower black eye-mask than Lesser Grey, with narrow white stripe above eye. Underparts white with faint grey wash on flanks. In flight, shows small white wing patch; broad white tail-sides. In juvenile, black areas greyer; thin white wing-bar. Call a soft trill. Kills small vertebrates and insects; impales prey for later consumption. Has bounding flight, also hovers.

Where to see Boggy fields with scrub, moorland, heaths. Scarce winter visitor to northern and central Italy.

Lesser Grey Shrike *Lanius minor* 20cm

Smaller and darker than Great Grey Shrike. Adult has grey crown and black mask (reaching forehead) with no white between. Back grey, wings black, underparts white with strong peach or rosy flush on breast and belly. In flight, shows large white wing patch and white tail-sides. Juvenile has white scaling on upperparts, slight peach flush on underparts. Quiet but may give harsh calls. Often seen perched on overhead wires.

Where to see Open but well-vegetated habitats; cultivated land. Scarce, local and declining summer visitor.

Woodchat Shrike *Lanius senator* 18cm

♂

♀

Stocky, boldly marked shrike. Male has chestnut crown, black face mask. Wings black with white patches, tail black, rump white. Underparts white, flushed light buff. Female has faint scaly barring on flanks, markings less clean-cut than male's. In flight, shows white sides to back, white comma marking at wing-bend, and white rump, tail-sides and tail-tip. Juvenile light grey-brown with darker scaly markings.

Song loud, warbler-like with squeaks, grating notes and mimicry, calls short, harsh. Typical shrike behaviour. Often seen perched atop bushes.

Where to see Habitat varied; open countryside with some trees, some bare ground. Scarce, local and declining summer visitor, commoner in south and on Sardinia.

Jay *Garrulus glandarius* 33cm

Colourful small crow. Mainly pinkish-peach, darker and greyer on back. Wing black and white, with bright blue, barred patch at wing-bend. Tail black, rump white. Black stripe below cheek, fine black streaks on crown. Eyes pale, bill black, legs pinkish. Round-winged in flight, very direct with steady wingbeats, white rump prominent. Usual call a harsh screech; sometimes imitates various birds of prey. Often very shy and quick to startle. Stores numerous acorns in autumn.

Where to see Woodland. Common, widespread resident.

Magpie *Pica pica* 45cm

Unmistakeable, with very long tail. Plumage black, glossed green and violet, with white belly (not reaching legs) and large white patch in wing. In flight, broad-winged, shows white outer flight feathers with black edges; graduated tail shape. Gives various rattling cackles and harsh or whining squawks. Takes varied diet, searching for food on ground with strutting walk or powerful hops. Often in small groups.

Where to see Nests in trees, found in all kinds of lowland habitats. A common, widespread resident.

Nutcracker *Nucifraga caryocatactes* 33cm

Distinctive small crow with front-heavy proportions; short tail, long, heavy bill. Plumage brown with dense large white spots, wings and crown uniform dark brown. Undertail and tail-tip white. Eyes dark. Round-winged in flight, white tail markings conspicuous. Quiet, but sometimes gives a rattling *krrrrrr* call. A specialist feeder on pine seeds, opening cones by gripping with foot while probing and prising cone open with its bill; also eats other seeds. Takes insects and other animal prey in spring and summer.

Where to see Upland coniferous forest. In Italy restricted to Alps, where common resident, but occasionally 'irrupts' further south in winter.

Alpine Chough *Pyrrhocorax graculus* 37cm

Smaller and shorter-billed than Chough, with shorter legs but longer tail. Adult and juvenile both have yellow bill, with barely noticeable down-curve. In flight, tail base looks 'pinched in', wings less deeply 'fingered' than in Chough. Usual call very different to Chough's, a high buzzing *zihhhr*. Gregarious, often in very large flocks that fly in compact formation, dipping, diving and rising in unison. Feeds on ground on invertebrates in summer, switching to plant matter in winter; also scavenges around ski lodges and other habitation.

Where to see A bird of high mountains (above c.1,500m), nesting on rocky ledges. Resident in Alps and Apennines.

Red-billed Chough *Pyrrhocorax pyrrhocorax* 39cm

All black with long, downcurved red bill and pink legs. Adult glossy black, juvenile dull black with shorter, yellowish bill. In flight, shows very long 'fingered' primaries; underwing shows dark inner wing contrasting with translucent flight feathers. Call a piercing, harsh, whirring *chiahh*. Agile and aerobatic in flight. Gregarious, usually seen in family groups. Feeds on ground, probing for worms. Nests on rocky ledge, sometimes on buildings.

Where to see Open, rugged upland habitats. Resident in western Alps and Apennines, rare and localised in Sardinia and Sicily.

Jackdaw *Corvus monedula* 32cm

Compact small, black crow. Adult has silvery-grey neck cowl, contrasting with black face, otherwise black. Juvenile completely black. Eye white (blue in juvenile), bill small and thick, black. Looks blunt-headed in fast, agile flight. Usual call a loud, bright, chuckling *tchack*, also downslurred *kyarrh*. Gregarious, often seen flying in large numbers. Feeds mainly on ground, omnivorous.

Where to see Forages in farmland and other open habitats. Nests (often colonially) on rocky ledges, on buildings or in hollow trees. Common, widespread resident.

Rook *Corvus frugilegus* 45cm

Black crow with longish straight bill. Has shaggy 'trousers' and peaked crown. Plumage has blue gloss. Adult has bare white bill base reaching to eyes; juvenile fully feathered on face and best told from Carrion Crow by head and bill shape. In flight shows round-ended tail. Calls various harsh caws. Gregarious, feeding in flocks and nesting in colonies (rookeries) in treetops. Joins foraging flocks of Jackdaws, eating worms, insects and plant material.

Where to see Farmland with mature trees. Winter visitor to northern Italy.

Carrion Crow *Corvus corone* 48cm

The northern counterpart of Hooded Crow, but all black. Smaller and more compact than Raven; sleeker than Rook with more rounded head and more curved upper mandible to bill. In flight, less agile than smaller Jackdaw or larger Raven; flies in direct line with steady 'rowing' wingbeats. Calls include cawing and squeaking or ringing notes. Adaptable omnivore; will scavenge carrion.

Where to see Forages on ploughed fields, rubbish dumps and other open habitats. A common resident in Alps, where readily hybridises with Hooded Crow.

Hooded Crow *Corvus cornix* 48cm

Largish black-and-grey crow, not readily confusable with any other species. Body plumage dull grey-brown; head, wings and tail black. Black from head spreads down into flaring patch on breast. Hybrids with Carrion Crow show more extensive black. Bill black, legs dark. Voice as Carrion Crow. Direct flight with deep wingbeats. Not especially gregarious, though groups may gather to feed. Searches ground for worms and insects. Usually shy and wary. Builds large stick nest well hidden in tree.

Where to see Common open-country resident throughout Italy.

Raven *Corvus corax* 60cm

The largest crow. All black, with powerful build, very strong bill with marked downward curve on upper mandible, head looks small relative to bill. Throat feathers often have puffed-out, shaggy appearance. In flight, shows diamond-shaped tail. Very agile and aerobatic, performing spectacular tumbles and turns. Voice a deep, resonant, rolling croak. Often seen in pairs patrolling territory; will chase birds of prey. Omnivorous, very attracted to carrion, can also kill quite large prey such as rabbits. Nests on a rocky crag or in a tree; nest very large.

Where to see Scarce and local in remote montane areas, craggy coasts and countryside with rocky outcrops.

Coal Tit *Periparus ater* 11cm

Small, large-headed tit. Has black cap and large flared black bib, white cheeks and white stripe on nape. Upperparts olive-grey with double white wing-bar; underparts buffish. Juvenile duller, tinged yellowish. Has ringing single-note calls, and song is usually a fast, repeated two-note phrase. A typical, very active tit, feeding mainly in treetops and often hovering or dangling from thin twigs.

Where to see Mainly in pine forest. Fairly widespread resident, mostly found in upland parts of mainland interior.

Crested Tit *Lophophanes cristatus* 11cm

Small tit with pointed crest. Upperparts warm mid-brown, underparts paler with buff wash. Head white, with black throat and narrow black stripe encircling neck; also has dark eye-stripe joining curved cheek-stripe behind eye. Crest marked with fine black-and-white barring, comes to tall point above centre of crown when bird is excited; otherwise held more flattened. Eyes dark red. Call a very high-pitched bubbly, cheerful trill; song combines trills and short, sharp notes. Feeds mainly in high treetops, picking insect prey from fine twigs. Flocks with other tits in winter.

Where to see Found in woodland (especially coniferous). Resident in Alps and north Apennines (increasing).

Marsh Tit *Poecile palustris* 12cm

Best distinguished from Willow Tit by call. Also has white spot at base of upper mandible (Willow has all-black bill); and cheek is two-toned with dividing line between white at front of face and light grey-brown wash behind. Lacks wing-panel. Has more balanced proportions compared with somewhat big-headed, thick-necked and 'egg-shaped' Willow. Call a sneezing *pit-choo*; song a repeated single- or two-note phrase.

Where to see Found in deciduous woodland, sometimes parks and gardens. Widespread, fairly common resident, not on Sardinia.

Willow Tit *Poecile montanus* 12cm

A big-headed, black-capped tit. Upperparts light grey-brown, with faint paler wing panel. Underparts white, shading to buff on flanks and undertail. Cheeks whitish, cap black, small black bib. Very like Marsh Tit; best separated by call, a repeated drawn-out nasal *djerrrr*. Song a series of sweet, rather slow notes; also a faster trill. Feeds at all levels in trees, and excavates nesting hole in soft decaying wood.

Where to see Mainly pine forest in upland, hilly countryside. Localised resident in Alps and Apennines.

Blue Tit *Cyanistes caeruleus* 11cm

Colourful small tit. Underparts yellow with narrow dark belly-stripe. Greenish back, shading to blue on wings and tail; narrow white wing-bar. Face white with blue crown patch outlined in white, dark bib, dark eye-stripes and cheek-stripes meeting on nape. Juvenile duller, with yellowish cheeks. Call a ringing three-note phrase; also churring alarm call. Song like extended call, last note a trill. Active, agile, primarily feeds in treetops.

Where to see Woodland, parks, gardens and other habitats with trees. Very common, widespread resident.

Penduline Tit *Remiz pendulinus* 11cm

juv.

Very small tit-like bird with shortish tail. Head light grey with black eye-mask (more extensive in male). Back reddish-brown, underparts light peach, with some red-brown spotting on breast forming upper breast-band in male. Wings well marked with dark feather centres and pale fringes, likewise tail. Juvenile has plain brown head. Bill grey, slim and sharply pointed, legs black. Call a soft downslurred *tsiu*; song includes repeated calls interspersed with high trills. Attaches its large, hanging sack-shaped nest to fine twigs; sometimes loosely colonial. Insectivorous.

Where to see Prefers damp marshy habitats and watercourses with trees. A localised and declining resident, also more widespread passage migrant.

♂

Great Tit *Parus major* 14cm

Largest tit species; colourful. Head black, cheeks white. Underparts yellow with broad black stripe from chin to belly (narrower in female). Back green, wings and tail bluish, narrow white wing-bar. Juvenile duller, with yellowish cheeks. Very varied calls include bright Chaffinch-like *twink*. Song a repeated two-note phrase, recalling a squeaky gate. Feeds at all levels in trees, taking insects, berries and seeds. Dominates other tit species in mixed feeding flocks.

Where to see Found in woodlands, parks and gardens. A very common, widespread resident.

Short-toed Lark *Calandrella brachydactyla* 15cm

Pale, rather thick-billed. Sandy-coloured on upperside with delicate black streaking, dark feather centres to wing feathers give strong wing pattern. Underparts almost unmarked white; slight fine streaking on breast-sides. Face pale with sandy cheeks and crown. Bill finch-like, yellowish. Legs pale pinkish. Call a short dry *trilp*, recalling House Martin; song comprises brief, chirruping phrases, given in undulating song-flight. Forages unobtrusively on ground.

Where to see Dry, sparsely vegetated open habitats, including coastal areas. Patchily distributed uncommon summer visitor, and more widespread passage migrant.

Woodlark *Lullula arborea* 14cm

Small, short-tailed lark. Well marked with brown upperparts streaked darker and paler, white underparts with buff flush on breast-sides, black streaking on breast. Distinctive black-and-white patch on wing edge. Pale supercilia meet in downward point on nape. Crown brown streaked black, cheeks unmarked reddish-brown with pale surround. Bill slim, longish, legs pale pink. Shows pale wing-bar and tail-corners in flight. Call a fluty, repeated *two-we*, song melodious, slow and melancholic.

Where to see Dry scrubby habitats; heathland, farmland, open woodland. Fairly common and widespread resident.

Calandra Lark *Melanocorypha calandra* 19cm

A large, stocky, big-billed lark. Light sandy-brown upperparts with darker brown streaking on crown, neck, back and wings. Has pale supercilium and eye-surround, darker cheeks. Black patches on breast-sides, breast otherwise only lightly streaked. In flight, shows conspicuous white trailing edge to wing, and solid dark (almost black) underwing. Bill yellowish, darker on top and at tip. Legs short, pink. Call a buzzing dry trill, song a long series of chirping and trilling notes. Singing birds rise to 100m or so and 'hang' with slow wingbeats.

Where to see Open farmland and grassland. Rather uncommon resident in southern Italy (including Sicily and Sardinia).

Skylark *Alauda arvensis* 17cm

Largish lark, proportionately small-headed, often shows crest. Upperparts light brown with dark streaks. Underparts pale, washed buff on breast with dark streaking. Brown crown and cheeks, face otherwise pale. Bill slimmish, legs pink. In flight, shows pale wing edge. Juvenile has scaly upperparts; no crest. Call a dry, rolling *chirrup*. Song a series of chirps, whistles and trills, given in towering song-flight. Forms loose flocks in winter.

Where to see Open farmland, grassland, alpine meadows and pasture. Common, widespread resident; numbers boosted by winter migrants.

Crested Lark *Galerida cristata* 18cm

A little larger than Skylark, shorter-tailed, with obvious spiky crest. Upperparts mid-brown with darker brown streaks and feather centres. Underparts whitish with dark streaking on breast, becoming finer on flanks. Face plain brown but with dark line running down from eye. Bill slim, longish, slightly down-curved. Legs pink, hind-claw not elongated. Broad-winged and short-tailed in flight; shows rusty underwing. Call a melancholic two- to four-note whistle. Song very varied, tuneful, given on ground or in flight.

Where to see Dry open habitats including farmland, waste ground, coastal flats, open fields. A fairly common resident in lowland areas.

Bearded Reedling *Panurus biarmicus* 15cm

♀

Very long-tailed. Plumage fawn-brown with black-and-white patches in wings. Male has grey head with black 'moustache' at base of bill; female has plain brown head. Juvenile more orange than adult, with black back, lores and tail-sides. Bill small, yellow; eyes pale, legs black. Call a dry, metallic *ping*. Song a simple squeaky phrase. Climbs among reed stems; takes insects in summer, reed seeds in winter. Usually in family groups.

Where to see Wetlands with reedbeds. Rare, local and declining resident, wandering somewhat in winter.

♂

Zitting Cisticola *Cisticola juncidis* 10cm

Tiny warbler-like bird with short but broad tail. Light sandy-brown upperparts with fine dark streaks on crown, cheeks and rump (crown nearly solid dark in male), and prominent broad dark streaks on back. Underparts unmarked, white with buff wash on breast-sides. Pale eyes, pinkish bill and legs. In flight, shows prominent white tail-tip; looks very round-winged. Call *chip*. Male sings in circling, undulating display flight over territory, fanning tail and giving single short, harsh *zit* calls at regular intervals.

Where to see Found in mostly lowland grassland with tall grass and some shrubs. A fairly common and widespread resident. Very hard and prolonged winters cause local temporary extinctions.

Icterine Warbler *Hippolais icterina* 12cm

Very similar to Melodious Warbler, replacing it in eastern Europe. Has longer wings and more strongly peaked head shape. Tends to be a little more richly coloured with more contrasting tones, and shows a faint pale panel in folded secondaries. Call an abrupt three-syllable note. Song (unlikely to be heard in Italy) varied, incorporating mimicry.

Where to see Woodland edges, scrubland, parks and gardens. A fairly common passage migrant.

Melodious Warbler *Hippolais polyglotta* 12cm

Robust yellow-green warbler with proportionately large head. Upperparts unmarked mossy green; underparts yellowish fading to whitish on belly. Face rather plain; dark eyes prominent. Crown often looks peaked. Bill strong, with pink base. Legs dull brownish. Has clacking or ticking calls. Song loud series of tuneful and harsher notes, somewhat similar to Sedge Warbler; may include mimicry; often sings in full view.

Where to see Open dry woodland, farmland with hedgerows, vineyards, scrub. Common, widespread summer visitor; passage migrant on Sicily and Sardinia.

Moustached Warbler *Acrocephalus melanopogon* 13cm

Has broad, round-ended tail and short wings. Upperparts warm reddish-brown, with almost black crown and fine black streaks on back and wings. Dark eye-stripes and cheeks, white throat and supercilia. Underparts strongly washed reddish-brown, fading to whitish on belly. Legs dark. Call low *chek*, song protracted series of *chrr chrr* and similar notes, includes drawn-out high whistles. Often feeds close to water's edge.

Where to see Reedbeds and other thick waterside vegetation. Scarce and localised resident in north-west and north-east; more widespread in winter.

Sedge Warbler *Acrocephalus schoenobaenus* 12cm

juv.

Paler than Moustached Warbler. Mid-brown above with dark streaks, crown darker. Broad creamy supercilia, dark eye-stripes, cheeks light brown shading into whitish chin. Underparts pale with buff wash on breast-sides. Juvenile has a little streaking on breast-sides. Legs dull pinkish-brown. Gives short sharp calls and low, rolling notes. Song a continuous, excitable series of chirrups, squeaks, dry grating sounds, sometimes given in short song-flight.

Where to see Wetlands and adjacent scrub. Passage migrant in Italy; a few breed in north-west.

Marsh Warbler *Acrocephalus palustris* 14cm

Almost identical to Reed Warbler. Dull olive-brown on upperparts, wing feather centres a shade darker. Paler on underparts with white throat. Slight pale eye-ring and supercilium in front of eye. Bill relatively heavy with pink base. Legs yellowish-pink. Calls mostly short, quiet notes. Song remarkable – tuneful, varied and including great variety of mimicry of other birds' calls and songs. Skulking and difficult to see.

Where to see Thick vegetation in wetland areas. Fairly common summer visitor in north and north-west.

Reed Warbler *Acrocephalus scirpaceus* 13cm

Very similar to Marsh Warbler but much more common in Italy. Dull mid-brown upperparts. Paler on underparts with definite white throat. Slight pale eye-ring and supercilium in front of eye. Bill long and slim. Legs dull brownish-pink. Calls soft *che* or *chik*. Song a continuous, steady-paced series of slightly nasal chirruping notes. Skulking and difficult to see, staying low in reeds, though may climb higher while singing.

Where to see Reedbeds; rarely seen far from waterside. Widespread, common summer visitor.

Great Reed Warbler *Acrocephalus arundinaceus* 18cm

Very large warbler; powerful well-proportioned build. As name suggests, recalls super-sized Reed Warbler. Upperparts mid-brown, underparts paler with white throat and belly, yellow-buff flanks and breast. Dark eye-stripe and narrow pale supercilium. Bill long and heavy, dark with pink base. Eyes mid-brown. Legs greyish, sturdy. Call a throaty *chack*. Song like Reed Warbler's but much louder and stronger, and includes more high-pitched notes. Less skulking than Reed Warbler; often sings in full view.

Where to see Found in extensive dense reedbeds in marshy areas, sometimes also in smaller stands of reeds alongside waterways. A fairly common and widespread summer visitor.

Savi's Warbler *Locustella luscinioides* 14cm

Plain, stocky warbler with long, broad, round-ended tail. Upperparts plain warm grey-brown, underparts paler, with whitish throat and centre of belly. Face almost unmarked; slight dark eye-stripe and paler supercilium. Undertail-coverts reach close to tip of tail, giving very thick look to tail base. Legs pinkish. Call a sharp *tzit*; song a continuous dry purring, like a fishing reel being turned. An inconspicuous, skulking bird, usually difficult to observe; climbs through vegetation almost like a small rodent.

Where to see Found in reedbeds and other dense waterside vegetation. A localised summer visitor to wetlands in northern Italy; may be found elsewhere on migration.

Grasshopper Warbler *Locustella naevia* 12cm

A small, very skulking mouse-like warbler. Has broad, round-ended tail. Plumage olive-brown, paler on underside, with strong dark streaking on crown, back, rump, flanks and undertail. Legs and bill base pinkish. Moves very discreetly through low vegetation, making occasional short flights. Song a continuous purr, like the sound of a cricket or the winding of an angler's reel, softer and more continuous than song of Savi's.

Where to see Scrubby wetlands and other habitats with bushes and dense ground cover. Passage migrant, more frequently seen in northern Italy.

Crag Martin *Ptyonoprogne rupestris* 14cm

A robust, drab martin. Upperparts mid grey-brown. Underparts a shade paler. Has faint dark streaks on throat and mottling on undertail. In flight, shows white patches in tail feathers, close to tips, and prominent blackish underwing-coverts. Tail has shallow notch (disappearing when tail fanned). Looks stockier and broader-winged than other martins, though still agile in pursuit of flying insects. Has various short, somewhat nondescript calls; song a quiet rapid twittering.

Where to see Nests in crevices in cliffs, also in walls of old buildings or under bridges, often in upland areas. In Italy found mainly in montane regions; migrants from further north move through on passage.

Sand Martin *Riparia riparia* 12cm

Looks slim and insubstantial in flight. Upperparts mid-brown; lacks white rump patch. Underparts whitish, with clear-cut brown breast-band. White throat patch curves up behind cheek, there fading into brown. Tail has shallow fork. Voice very dry, grating; gives short single notes and longer series when at nest. Colonial, tunnel-nesting breeder.

Where to see Nests in soft earth banks; often next to water but also in quarries. Mostly hunts over open water. A fairly widespread summer visitor, and very widespread passage migrant.

Barn Swallow *Hirundo rustica* 19cm

Long-tailed, highly aerial bird. Upperparts blackish with bright violet-blue gloss. Has dark orange forehead patch and throat, dark breast-band. Underparts whitish with variable peachy tint. Outer tail feathers narrow, long; tail shows white spots near feather bases when fanned. Juvenile more muted, tail shorter, lacking elongated outer tail feathers. Has pleasant twittering calls, and a sharper *vit* note, extended to *si-vit* when alarmed.

Where to see Nests on ledges inside buildings; hunts over meadows, farmland and open water. Common, widespread summer visitor.

Red-rumped Swallow *Cecropis daurica* 17cm

Slightly smaller than Barn Swallow, with thicker tail-streamers. Tail, wings and crown dark, glossed blue. Nape and top of rump light orange, lower rump whitish. Face and most of underparts light fawn with fine darker streaks; undertail dark. Tail lacks white markings. Calls more drawn-out and rasping than those of Barn Swallow. Builds funnel-shaped mud nest on crag or building.

Where to see Hunts over water and open countryside, often nests on bridges over water. Rare and localised summer visitor in central and southern Italy.

House Martin *Delichon urbicum* 14cm

Upperparts, including upper cheeks, black with blue gloss; has square white rump patch. Underparts white, including legs (the only songbird with feathered legs and feet). Bill small, dark. Juvenile duller grey-black; white areas suffused dusky grey. Wings broad-based with pointed tips, tail forked. Calls and song are dry rolling twitters. Hunts for flying insect prey. Builds mud cup nest.

Where to see Nests in colonies on buildings or on cliff faces; hunts over open country and water. Common, widespread summer visitor.

Wood Warbler *Phylloscopus sibilatrix* 12cm

Brightly coloured, rather large with big head, short tail and long wings. Upperparts quite bright mid-green; flight feathers blackish with white fringes. Face, throat and upper breast bright yellow; has dark eye-stripe with yellow supercilium. Belly clean silky white. Legs dull greenish. Call a short *zip*. Song includes shivering trills and slower, purer *tyuh* notes. Sings and forages in treetops but nests close to or on ground.

Where to see Deciduous woodland. Summer visitor, most common in central upland areas, passage migrant elsewhere.

Yellow-browed Warbler *Phylloscopus inornatus* 10cm

Tiny, colourful leaf warbler with small bill and short tail. Green upperparts with strong dark eye-stripes, yellow supercilia and yellow double wing-bars, flight feathers have whitish tips. Underparts light greyish, legs fairly dark. Call a drawn-out *tscheet*. Very agile and active, recalling Goldcrest or Firecrest; feeds in foliage and frequently hovers while picking insect prey from outermost leaves. Often approachable and may flock with other small birds.

Where to see Scrubby and lightly wooded areas. Uncommon passage migrant in mainland Italy but common in autumn on small islands, especially Pelagie Islands south of Sicily.

Chiffchaff *Phylloscopus collybita* 11cm

Very like Willow Warbler but a little drabber and less sleek-looking, with shorter wings. Upperparts mossy olive-green, underparts paler. Has less bold eye-stripes and supercilia than Willow Warbler, and darker cheeks against which white under-eye crescents (half eye-rings) are prominent. Legs usually darkish. Call a short upslurred *hweet*, or downslurred *pyoo*. Song repeated steady *chiff-chaff-chiff-chaff*. Usually feeds and sings high in trees, but nests low down.

Where to see Woodland, scrub, parkland, larger gardens, wooded riversides. Common, widespread resident.

Willow Warbler *Phylloscopus trochilus* 12cm

Slimmer and duller than Wood Warbler. Upperparts mossy olive-green, underparts whitish with yellow-green tints, fading to almost white on lower belly. Dark eye-stripes and pale supercilia, cheeks also rather pale with faint darker outline. Legs usually pinkish. Juvenile brighter, with yellow on entire underside. Call a two-note *hoo-weet*. Song a sweet steady series of descending notes.

Where to see Woodlands, parks, gardens and other habitats with at least some trees. A northern European summer visitor, occurring as common passage migrant in Italy.

Western Bonelli's Warbler *Phylloscopus bonelli* 11cm

A small, plain-faced leaf warbler. Crown and back dull grey-green, but wings, rump and tail much brighter green (with dark feather centres to wings and tail). Face greyish with full pale eye-rings; dark eyes prominent. Shows a hint of dark eye-stripes and paler supercilia. Underparts white. Bill shortish and strong-looking with pink base. Legs dull greyish. Call a two-note, upslurred *hu-wee*; song a repeated high-pitched fine note. Forages in high treetops, often hovering to pick insects from leaves.

Where to see Prefers dry woodland, mainly oak and pine, on hills and mountains. A summer visitor mainly to upland areas of central and northern Italy; passage migrant elsewhere.

Long-tailed Tit *Aegithalos caudatus* 14cm

Tiny short-billed, neckless tit-like bird, tail about 50% of total length. Whitish with blackish back, tail, wings and stripe above eye. Shoulders, flanks and belly washed rosy-pink. Dark eye has orange eye-ring. Bill and legs dark. Juvenile lacks pink tones and black head stripe extends to whole upper face; eye-ring red. Call a short purring *prrrrt*, also shorter *sst* notes. Social; at most times of year seen in family parties which follow one another from tree to tree. Builds beautiful lichen-covered ball-shaped nest well hidden in bush.

Where to see Found in woods, parks, gardens, scrubland and similar habitats. Common resident (not Sardinia).

Cetti's Warbler *Cettia cetti* 13cm

Dark, with long, rounded tail. Upperparts warm, rich red-brown, underparts slightly paler grey-brown, barely paler than upperparts. Faint darker mottling on undertail. Has slight pale supercilia, dark eye-stripes. Wings very short, round-ended; tail has 'heavy' look but is often held cocked. Calls loud, rich with fluty quality. Song short, loud, explosive series of very fluty notes. Very skulking, creeping unobtrusively like rodent, though males in spring occasionally sing in the open.

Where to see Bushes and scrub close to water. Common, widespread resident.

Blackcap *Sylvia atricapilla* 14cm

♂

♀

A sleek grey warbler. Upperparts unmarked mid grey-brown, underparts a shade lighter. Male has black cap, female and juvenile reddish-brown cap (and slightly more brown-toned body plumage). White under-eye crescent, whitish undertail and throat. Cheeks grey rather than whitish as in Marsh and Willow Tits. Bill strong for warbler, legs grey. Call a hard *tac*, given repeatedly when alarmed. Song loud, very sweet and melodious short fluting phrases; also has softer, more mumbled and continuous subsong. Forages unobtrusively in foliage.

Where to see Found in woodland, parks, gardens, scrub, hedgerows and riversides with trees. Common, widespread resident.

Garden Warbler *Sylvia borin* 14cm

Sturdy, plain-looking. Upperparts dull grey-brown with subtle pure grey wash on neck-sides. Underparts paler; whitish undertail but strongly washed buff-brown on breast-sides and flanks. Slight hint of eye-stripe and paler supercilium; dark eye prominent. Bill rather short and stout. Legs dark grey. Call a slightly throaty *chek*. Song recalls Blackcap's subsong; melodious but hurried, in long phrases. Inconspicuous and best located by song.

Where to see Woodland, scrub, large parks, woodland edges. A summer visitor to north and north-east Italy; passage migrant elsewhere.

Barred Warbler *Curruca nisoria* 15cm

A large and stocky, strong-billed warbler. Adult light smoky grey, paler on underside, with scaly barring from throat to undertail and whitish fringes to wing feathers. Legs and bill greyish, striking yellow eyes. Juvenile (more likely to be seen in Italy) rather plain and uniform grey with only slight, faint barring on flanks and undertail, wing feathers with pale fringing, eyes dark. Call a grating *arrrr-at-at*. Feeds on berries and insects.

Where to see Scrubby and lightly wooded habitats. Uncommon passage migrant, a few pairs still breed in far north-east.

Lesser Whitethroat *Curruca curruca* 12cm

Small, rather short-tailed warbler. Grey crown, darker cheeks, giving hint of face mask. Upperparts otherwise dull grey-brown. Underside pale with brown-washed flanks, white throat and undertail. Eyes dark with slight white 'spectacle' markings; legs dark grey. Call a soft, clicking *tet* or repeated churring scold. Song simple short rattle, like Cirl Bunting; also soft fast warbling. Often sings from within cover.

Where to see Scrub with scattered trees, woodland edges, usually near or above tree-limit. Fairly common summer visitor in Alps; passage migrant elsewhere.

Western Orphean Warbler *Curruca hortensis* 15cm

A large, long-tailed warbler. Male's crown, cheeks and tail black, wings and back soft mid-grey, underparts from throat to undertail white with light grey-pink wash on flanks and breast-sides; eyes pale. Female similar but browner and drabber with dark eye-mask but paler brownish crown, juvenile like female but eyes darker. Legs dark, bill grey with black tip. Has simple four-syllable song, calls include hard *chak* and rattling *drrr*.

Where to see Montane hedgerows, scrub and similar habitats. Uncommon passage migrant, rare and sparsely distributed summer visitor.

Sardinian Warbler *Curruca melanocephala* 13cm

Slim, long-tailed, relatively large-headed with prominent red eye-rings. Male dark grey, white on underparts with grey-washed flanks. Shows white tips to tail feathers when tail fanned. Female recalls dark male Common Whitethroat, with grey head and dark grey-brown body, tinged pinkish on underparts, whitish throat, belly and undertail, red eye-rings. Call single *tschak* or fast, hard rattle; song short, scratchy chatter.

Where to see Warm open woodland, tall scrub, parks and gardens. Common resident, but rare and localised in far north.

Eastern Subalpine Warbler *Curruca cantillans* 12cm

Small, colourful, short-tailed. Male upperparts dark blue-grey, shading to browner on wings. Throat and breast deep brick-red, gradually shading to pale pink on flanks, whitish on belly. Has prominent white stripe separating grey cheek from red throat; prominent red eye-ring. Legs pale pinkish. Female much paler; juvenile plain brown but with dark-centred wing feathers. Call dry *chak*, song muddled series of twitters, trills and dry rattles.

Where to see Scrubland, woodland edges. Fairly widespread summer visitor south of Tuscany; on Sardinia, passage migrant only.

Moltoni's Warbler *Curruca subalpina* 12cm

Very like Eastern Subalpine Warbler. Upperparts mid blue-grey, underparts soft pinkish-grey, only slightly paler on belly. White line between grey cheek and pink throat is narrow. Red eye-rings; pinkish legs. Female is washed-out version of male; juvenile similar to juvenile Eastern Subalpine Warbler. Call a dry, hard rattle. Song is similar to Subalpine's but distinctly faster and more frantic, often with hissing tone.

Where to see Breeds from sea level to low mountains in scrubland, woodland edges and other open habitats with bushes. Summer visitor, arriving slightly later than Subalpine Warbler; found in north-west Italy and on Sardinia; scarce passage migrant down mainland east coast.

♀

imm.

♂

Common Whitethroat *Curruca communis* 14cm

Slim with relatively big head and long tail, rufous wings. Male has grey head with white eye-rings, grey-brown back and tail. Underside pale, washed pinkish, throat bright white. Eyes light brown, legs pinkish. Female similar but browner on head. Juvenile like female, but eyes dark. Call a scolding *djerr*. Song fast, rather dry scratchy warble, often given in short, steeply rising song-flight. Often conspicuous.

Where to see Bushy grassland, farmland with hedgerows, heaths, and other scrubby countryside. Common and widespread summer visitor.

Marmora's Warbler *Curruca sarda* 13cm

Small, slender, very distinctive dark warbler with long tail and large head, crown often peaked and throat puffed out. Male uniform mid slate-grey, with dark throat but slightly paler undertail. Legs pinkish-orange, bill yellowish with dark tip, eye light brown with red eye-ring, lores blackish. Female very slightly browner grey, lacks dark lores so eye-ring less striking. Juvenile similar, tinged more brownish. Call a harsh, Stonechat-like *tchek*, song a rapid but sweet-toned warble.

Where to see Found in dense scrubland on hillsides and down to coast level. Resident on Sardinia and some other islands, also rare winter visitor to far south of mainland Italy and Sicily.

Spectacled Warbler *Curruca conspicillata* 12cm

Recalls petite, richly coloured Common Whitethroat. Male has blue-grey head, blackish around bill base, with bold white eye-rings. Back and tail grey-brown, wings rufous and shorter than Common Whitethroat's. Underparts white, strongly washed pinkish, throat white. Female like miniature, small-billed female Common Whitethroat. Call a hard rattling *trrr*, song a rapid, high-pitched twitter, sometimes preceded by slower, purer notes. Shyer than Common Whitethroat.

Where to see Scrubby open habitats, often in uplands. Summer visitor to southern Italy, scarcer and more localised further north.

Dartford Warbler *Curruca undata* 13cm

A dark warbler, very similar to Marmora's Warbler in shape and attitude but more colourful (though looks completely dark in poor light). Male is dark slate-grey on upperparts, with dark rich purplish-red underside from throat to undertail; bright red eye-ring. Throat marked with fine white speckles. Legs yellowish. Female slightly less richly coloured version of male; juvenile rather dull earth-brown with slightly paler throat. Call a protracted *churr*; song a short rapid warble, mostly dry-toned but with a few fluted notes.

Where to see Breeds in scrubby countryside; woodland edges, heaths, coastal bushes. Resident on mainland Italy, Sicily and Sardinia.

Red-billed Leiothrix *Leiothrix lutea* 15cm

A colourful songbird, resembling a small thrush, with striking red bill. Head greenish with pale patch around prominent dark eye, throat yellow shading to orange on breast, otherwise greyish with bright red-and-yellow wing markings. Legs light pinkish. Has rich, fluting song. Often occurs in small family parties, unobtrusive but not especially shy.

Where to see Wooded habitats, parks. Native to Asia, established in parts of Europe from escaped pets; has patchy but increasing populations in northern and central Italy.

Goldcrest *Regulus regulus* 9cm

Tiny, rounded, neckless bird with short tail. Plumage olive-green, paler below, with double white wing-bar. Top of crown yellow (orange-centred in male) with black edge. Eye prominent in plain, pale face, dark line extending under eye from bill base. Legs dull brownish, bill dark, shortish. Juvenile has plain grey-green crown. Call very high *tzee-tzee*, song a high-pitched twittered phrase, ending with louder flourish. Often hovers. Fearless, very active.

Where to see Woodland of all kinds. Breeds in Alps and Apennines, more widespread in winter elsewhere.

Firecrest *Regulus ignicapilla* 9cm

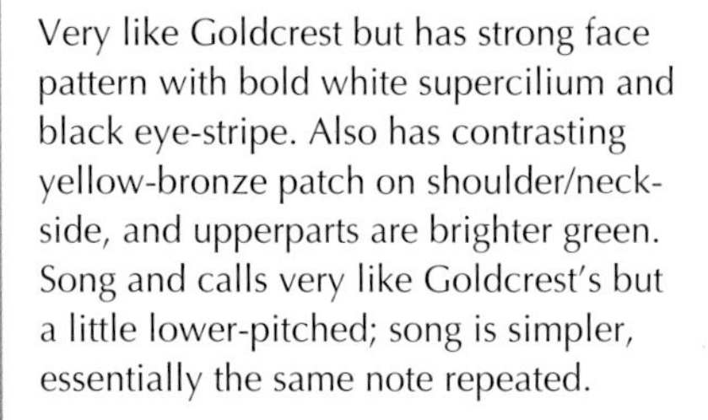

Very like Goldcrest but has strong face pattern with bold white supercilium and black eye-stripe. Also has contrasting yellow-bronze patch on shoulder/neck-side, and upperparts are brighter green. Song and calls very like Goldcrest's but a little lower-pitched; song is simpler, essentially the same note repeated.

More likely to forage at lower levels than Goldcrest, but not unusual for both species to be seen together in winter flocks.

Where to see Found in woodland but also more open scrubby countryside. Fairly common and widespread resident.

Wallcreeper *Tichodroma muraria* 16cm

Treecreeper-shaped but with longer, thinner bill. Blue-grey body plumage, becoming white on throat and breast. Wings have extensive deep pink-red patches. Breeding-plumaged male acquires black lower face and throat. Legs and bill black. In flight, shows white spots near wingtips, white tail-corners, and strikingly round, butterfly-like wing-shape. Song soft rising and falling whistles. Climbs on steep rock faces; probes cracks for prey.

Where to see Breeds on steep rockfaces in Alps and Apennines. In winter more widespread on lower ground in crags, quarries, ruins, small historic cities and even craggy coasts.

Nuthatch *Sitta europaea* 13cm

Agile tree-climber with large head, long strong bill and short tail. Upperparts from crown to tail mid blue-grey. Has long black eye-stripes, with white cheeks below, shading gradually to warm orange-buff. In male, colour darkens to almost brick-red on undertail, with contrasting white spots; remains paler in female. Legs pinkish, very sturdy with strongly curved claws. Call a loud, ringing *tuit* or *vit*; song of similar notes.

Where to see Woodlands. Common resident; local in Sicily and absent in Sardinia.

Treecreeper *Certhia familiaris* 13cm

Almost identical to Short-toed Treecreeper; best distinguished by voice, but also whiter on underside, and 'Z' pattern in folded flight feathers less clearly defined. Hindclaw slightly longer, bill slightly shorter. Call a very high, thin whistle, Goldcrest-like, with buzzy quality. Song short phrase, high and buzzy, falling in pitch and finishing with trill. Behaviour like Short-toed Treecreeper.

Where to see Upland mixed or pine forests. Common resident in Alps, scarcer and more localised in Apennines and south Italy; absent from Sicily and Sardinia.

Short-toed Treecreeper *Certhia brachydactyla* 13cm

Upperparts mid-brown with paler streaks, pale supercilium and cheek patch. 'Z'-shaped pattern of dark and light on folded flight feathers. Underparts white, shading to brownish on rear flanks. Tail long and narrow, feather tips pointed. Bill long and downcurved. Legs pinkish, short but with strong toes for climbing and clinging. Call a ringing *tuit;* song short, jerky phrase of similar-pitched notes. Climbs in spirals on tree trunks and branches.

Where to see Lowland woodlands, parks and gardens. Common and widespread resident; not Sardinia.

Wren *Troglodytes troglodytes* 10cm

Very small, rounded, short-tailed, with longish, slightly downcurved bill. Plumage brown, finely barred, paler and greyer on underparts. Dark eye-stripe, pale supercilium. Legs pinkish, rather long, feet strong. Usually holds tail cocked upright over back. Call a hard, dry rattle or single *clak* note; song very loud series of fast, hard rattling trills. Often forages on ground, clambering through dense vegetation, but will sing from raised, conspicuous perch.

Where to see Woodlands, gardens, hedgerows, scrubland. A very common, widespread resident.

Dipper *Cinclus cinclus* 19cm

Unique riverine songbird; stout and short-tailed, recalling oversized Wren in shape. Adult dark brown; throat and breast white with crisp edge. Juvenile very grey with prominent paler scaling, throat white, breast and belly barred grey. Flight call a sharp *tzit*; song a varied, slow-paced phrase of rather tuneless notes. Plunges into water, finding prey on riverbed. Flies fast and low along river; also floats downstream with current.

Where to see Fast-flowing rivers and streams with rocks to use as perches. Uncommon, widespread resident.

Starling *Sturnus vulgaris* 21cm

Dark, glossy bird with short tail, strong pointed bill, upright stance and bustling walk. In breeding season, uniform blackish glossed green and violet, a few paler spots. Bill yellow, legs pinkish. In winter similar but with numerous whitish spots; bill darker. Juvenile light brown with dark lores and pale throat; gradually develops spotted winter plumage. Various harsh calls; song includes buzzes, whistles and mimicry.

Where to see Farmland, parks, towns. Common resident in northern and central Italy, scarce in south; more widespread in winter.

Rose-coloured Starling *Pastor roseus* 21cm

Same size and shape as Starling; also similar in behaviour. Adult has rose-pink body, with black head, neck, upper breast, wings and tail; long shaggy crest, pink bill and legs. Juvenile markedly paler than juvenile Starling, with yellowish bill and dark feather centres on wings; juvenile plumage retained longer into autumn than in Starling.

Where to see Open habitats, often joining other starling flocks. Nomadic, may turn up anywhere, most likely in late spring (more abundant some years than others). Has bred in Italy in the past.

Spotless Starling *Sturnus unicolor* 21cm

Very like Starling but less glossy, with sheen toned purple and no green tones. Entirely unspotted in breeding plumage. In winter plumage has fine paler spots but much less conspicuously spotty than Starling. Legs lighter, brighter pink than Starling's. Juvenile darker than juvenile Starling, with entirely dark upper face. Voice similar to Starling's but with sharper, clearer tone to some notes. Behaviour similar to Starling's; mainly forages on ground, walking briskly and pausing to probe soil with bill. Gregarious.

Where to see Usually breeds and forages close to human habitation; farmland, parks and similar open ground. Resident only on Sicily and Sardinia.

Mistle Thrush *Turdus viscivorus* 28cm

The largest thrush; long-tailed, pot-bellied and small-headed. Upperparts light sandy-brown with conspicuous pale edges to wing feathers. Underparts pale with dense round spots over entire breast and belly. In flight, shows white inner part of underwing and pale tail corners. Call a hard dry rattle, very loud when mobbing predators; song a loud, powerful fluting but rather simple phrase.

Where to see Parks, gardens, woodland edges, mountainsides. Fairly common and widespread resident in mountains, scarcer in north and east.

Song Thrush *Turdus philomelos* 21cm

Smaller than Mistle Thrush, with compact proportions. Upperparts uniform warm mid-brown. Underside whitish with strong yellowish wash to breast-sides. Spots teardrop-shaped, becoming small and sparse on centre of belly. In flight, shows yellowish-orange inner part to underwing. Call soft *tic*. Song comprises short, fluting phrases, each repeated once or more. Feeds on ground and in bushes, taking invertebrates and berries.

Where to see Parks, gardens, woodland and farmland. Resident in central mainland Italy from north to south, winter visitor and passage migrant elsewhere.

Redwing *Turdus iliacus* 21cm

(Near Threatened) Upperparts dark brown with pale edges to wing feathers. Has prominent white supercilia and cheek-sides, dark eye-stripes. Underparts spots coalesce into streaks on flanks. Rear flanks and inner part of underwing bright rusty-red. Call grating *zeeeh*, noticeable at night as flocks migrate overhead. Song (sometimes heard in spring) varied sequence of whistled, fluty and squeaky notes. Roams in flocks with other thrushes.

Where to see Visits stands of berry-bearing trees and bushes, and forages on bare fields. Widespread but scarce winter visitor.

Blackbird *Turdus merula* 25cm

Sleek, long-tailed. Male matt black; legs dark, bill and eye-rings yellow. Female dark brown with variable paler throat and vague speckling on breast and belly; bill orange or blackish. Juvenile warmer brown with yellowish spotting. Call a liquid chuck or more agitated *quick-quick-quick*. Song strong fluted phrases at generally quite slow pace. Hops on ground, listening for prey; also takes berries and windfall fruit.

Where to see Woodland, parks, gardens and other habitats with bushes or trees and some open ground. Common, widespread resident.

Fieldfare *Turdus pilaris* 25cm

Rather colourful, large thrush. Head, lower back and rump light grey, upper back and shoulders dark reddish-brown, tail black, underparts white with strong orange wash on breast, marked with black streaks that become chevrons on flanks; central belly and inner part of underwing strikingly white. Has blackish lores and white supercilium. Call loud clucking chuckle *chek-chek*. Song a rather simple chatter. As with Redwing, forms roving winter flocks.

Where to see Any open countryside, especially around berry-bearing shrubs. Widespread winter visitor, scarcer in south.

Ring Ouzel *Turdus torquatus* 24cm

Smaller and greyer than Blackbird, with scaly pale feather fringes on underparts and wings, and prominent white throat gorget (suffused brown in female). Bill yellowish with black top and tip, legs dark, lacks prominent eye-rings. Calls similar to those of Blackbird, song less melodious with more chattering pace. Rather shy.

Where to see Breeds in rocky uplands, on passage visits scrubby and grassy habitats. Fairly common summer visitor in the Alps, scarce in central Appenines; may occur as passage migrant elsewhere.

Spotted Flycatcher *Muscicapa striata* 14cm

Sleek and slim with large head, short legs, long wings. Adult dull grey-brown on upperparts, with crown, breast and flanks whitish with brown streaks; belly and undertail whitish. Juvenile similar but has pale spotting on back and wings. Legs dark. Call short *see*, song short phrase of scratchy notes. Perches upright before flying to catch prey, often returning to same perch. Nests in tree holes or crevices in walls. In Sardinia, southern coastal Tuscany and nearby islands, the paler, shorter-winged and less well-marked subspecies *balearica* (Tyrrhenian Flycatcher) occurs.

Where to see Woodland with sunny clearings, parks and gardens. Fairly common, widespread summer visitor.

Robin *Erithacus rubecula* 13cm

Rotund, smallish chat. Adult has brown upperparts and orange-red face and breast, separated by variably broad greyish line. May show narrow paler wing-bar. Belly whitish. Eyes prominent, large and dark. Legs dark. Juvenile upperparts uniform warm mid-brown with paler spotting, underparts paler and spotted. Call sharp *tic*; song sweet, melancholic twitters and trills, including slow phrases. Often perches with tail raised, wings drooping. Usually forages on ground, takes invertebrates; also berries.

Where to see All kinds of habitats with trees and ground cover. Common resident, more widespread in winter.

Nightingale *Luscinia megarhynchos* 16cm

Recalls all-brown, long-tailed Robin. Plumage rather uniform warm mid-brown, with greyish wash on neck-sides, and slightly paler breast shading to whitish belly. Eyes dark with subtle paler eye-ring, legs pinkish. Tail has strong rufous tint. Frequently perches with tail raised and wings drooping. Song exceptional; powerfully throbbing, fluting phrases alternating with slow, high single-note whistles, given by day and through night. Shy and skulking.

Where to see Woodland with dense, tangled undergrowth, also scrubby hedgerows and large gardens. Common, widespread summer visitor.

Bluethroat *Cyanecula svecica* 14cm

Robin-like chat with strongly striped face pattern. Upperparts grey-brown, underparts whitish. Male in breeding plumage has vivid blue throat patch, which may bear a red or white central spot (depending on subspecies), with dark lower border and reddish band below that. Female has dark-edged pale throat. Juvenile boldly pale-spotted; first-winter like female but duskier. A shy bird. Call *chak*, song a clear, rapid stream of notes.

Where to see Light woodland, scrub, reedbeds. Fairly common passage migrant, especially near water, rare in winter. Has bred occasionally in the Alps.

Collared Flycatcher *Ficedula albicollis* 13cm

Very like Pied Flycatcher but with more extensive white. Male has large white forehead patch, and complete white collar. Rump white, with diffuse edges. White wing patch large, with adjoining broad white wing-bar. Female very like female Pied Flycatcher; white wing-bar broader. Call a thin whistle. Song a slow sequence of slightly rasping whistles, shifting considerably in pitch. Nests in tree holes (often old woodpecker nests) and feeds on insects caught in flight.

Where to see Breeds in open sunny deciduous woodlands, parks, gardens and similar habitats. Rather patchily distributed summer visitor to central and southern mainland Italy; passage migrant elsewhere.

Pied Flycatcher *Ficedula hypoleuca* 13cm

Compact, short-tailed flycatcher. Male has black upperparts with white forehead patch and large white wing patch, underparts from throat to undertail pure white. Female has similar pattern but black areas replaced by grey-brown; lacks forehead patch and has less white in wing than male. Legs black. Call a sharp *tik*; song a loud, rather unmelodic pulsing trill with slight changes in pitch.

♀

Where to see Open, sunny oak woodland; other kinds of wooded habitats on migration. Widespread passage migrant in Italy.

♂

Black Redstart *Phoenicurus ochruros* 14cm

imm.

Like Common Redstart in shape. Male sooty blackish, darkest on face and breast, fading paler on belly; often has paler forehead. Variably prominent white panel in wing. Female, juvenile and younger males uniform dull grey-brown, paler on belly. Tail and rump red in all plumages. Call a short whistle, song a simple twitter interspersed with curious grinding noises. Nests in crevice in rock or building.

Where to see Open, rocky countryside, also urban environments. Fairly widespread resident, with migrants from further north arriving in winter.

♂

Common Redstart *Phoenicurus phoenicurus* 14cm

Slender chat with long wings. Male colourful (especially in spring), with blue-grey crown, back and wings, white flash on forehead, black face and throat, orange-red breast and belly. Legs blackish. Female and juvenile rather plain warm brown, with reddish wash to breast and flanks. In all plumages has red rump and red, dark-centred tail which it frequently shivers. Call *tick-tick* and soft *tuit*; song pleasant descending twittered phrase.

Where to see Deciduous woodland, parks and large gardens. Fairly widespread summer visitor, widespread passage migrant.

Common Rock Thrush *Monticola saxatilis* 19cm

Large, stocky, short-tailed and long-winged thrush-like chat. Male colourful, with blue-grey head and upper back, deep brick-red breast and belly, dark wings, white lower back and red, dark-centred tail. Female uniform brown, a shade lighter and redder on the underside, with pale scaling on upperparts and dark barring on underparts; reddish tail. Warmer-toned than female Blue Rock Thrush. Legs dark. Call a hard *chat* or softer *whst*; song quiet, melodious with melancholy tone; often given in undulating, bat-like display flight.

Where to see Mainly found in montane areas above 1,500m, with steep cliff faces and areas of grassland. A localised summer visitor and passage migrant.

Blue Rock Thrush *Monticola solitarius* 22cm

Larger, longer-tailed than Common Rock Thrush, with distinctly longer bill. Male deep dark blue with darker wings and tail; looks completely black in all but good light. Female dark brown, uniform on upperside but paler below with strong dark barring; no rufous tones. Legs dark. Has a two-note call and a single low-pitched *chek*; song like Common Rock Thrush's but louder with quavering notes. Tends to be shy and flighty, though chooses conspicuous perches.

♂

♀

Where to see A bird of rocky habitats, usually at lower altitudes than Common Rock Thrush and found down to sea level, also in towns. Fairly common, widespread resident.

Stonechat *Saxicola rubicola* 12cm

♂ br.

♀

Slightly stouter and shorter-winged than Whinchat. Male has black head and blackish upperparts with dark cheeks, prominent white neck-sides, and orange underside down to belly. Rump pale. Narrow white shoulder patch. Female similar but paler, head brown, lacks white neck patches. Call *weet tak-tak*, with last two notes hard tapping sounds; song a brief, Dunnock-like twitter. Perches atop bush or post, dropping to ground to catch insects.

Where to see Open but well-vegetated heaths, alpine meadows, waste ground. Common and widespread resident.

Whinchat *Saxicola rubetra* 13cm

Rotund but relatively long-winged chat. In all plumages has long pale supercilium and white tail base. Male boldly marked with blackish crown and cheeks, white supercilium and cheek-sides, orange breast shading to white belly, and sandy-brown upperparts with strong dark streaks; small white wing patch. Female similar but head markings duller, more uniform. Calls are soft clicks and whistles, song short, varied, with harsh and more tuneful notes.

Where to see Wet scrubby meadows, moorland, forest edges. Rather localised summer visitor and widespread passage migrant.

Northern Wheatear *Oenanthe oenanthe* 15cm

Rather large, slim with upright posture, white rump and tail with black centre and tip. Male light blue-grey on crown and back, with black wings and eye-mask, white supercilium. Underside pale with pale peachy breast, whiter on lower belly. Female paler and browner with brownish eye-stripe. Juvenile dull grey with scaly markings, becoming more like female but very orange-toned after first moult. Whistling and *chak* calls, song rapid dry chirruping.

Where to see Open rocky habitats with short or sparse vegetation. Fairly widespread summer visitor.

♂

♀

Isabelline Wheatear *Oenanthe isabellina* 16cm

Resembles a large, long-legged and rather plain and pale female Northern Wheatear. Plumage uniform sandy brown, tinged peach on neck-sides, with dark lores, darker centres to wing feathers, white rump and mostly black tail (with just the bases of outer feathers white). Has a penetrating cheeping call, with a more clicking tone when alarmed. Behaviour similar to that of Northern Wheatear.

Where to see Open habitats with short or sparse vegetation, often coastal. Rare passage migrant in spring in Sicily and its small islands, still vagrant elsewhere.

Eastern Black-eared Wheatear *Oenanthe melanoleuca* 14cm

♀

Smaller and 'rounder' than Northern Wheatear. Male whitish with black face and wings (a few have pale throat); crown, back and breast-sides have sandy tone. Female rather uniform earth-brown with slightly paler underside, no strong face markings. Male in fresh autumn plumage much browner than spring male. In all plumages shows white rump and tail-sides; centre, tip and lower sides of tail black. Calls like Northern Wheatear's; also a hissing note. Song a fast, dry twittering, similar to Common Whitethroat, and may include mimicry.

Where to see Open, arid countryside with some bushes, riversides, sometimes around villages. A summer visitor mainly to south; very localised further north.

♂ pale-throated
♂

Red Avadavat *Amandava amandava* 9cm

Tiny, finch-like bird with distinctive tapering cigar shape owing to full feathering around tail base. Male bright red, shading to blackish on wings, belly and tail, with white speckles on wings and underparts, bill red, legs pinkish. Female mostly light grey but with dark lores, red rump, black tail and red bill. Flocks chirrup in flight; song a sweet, twittered phrase.

Where to see Wetlands and damp meadows with long grass. Established from cagebird escapees in parts of Tuscany and Basilicata.

Alpine Accentor *Prunella collaris* 16cm

Head grey, upperparts grey-brown streaked black with two thin white wing-bars either side of blackish wing panel. Underparts grey shading to rich red-brown. Patch of black-and-white barring on throat. Bill shortish, dark with yellow base. Call rolling *drrru* or hard *chak*; song (given by both sexes) combines trills and squeaky notes. Often in small groups. Feeds on insects and berries.

Where to see Sparsely vegetated high uplands; may move lower in winter. Resident in Alps and Apennines, more widespread in winter, even along rocky coasts.

Dunnock *Prunella modularis* 14cm

Small, inconspicuous, dark songbird. Head and upper breast dull grey with browner cheeks, body brown with dark streaks on underside. Female usually has browner, streakier head than male; juvenile spotted on head and breast. Song short, high-pitched warbled phrase. Call loud, plaintive single whistle. Feeds mainly on ground, constantly flicking wings and tail, sings from high perch. Takes insects in summer, seeds and berries in winter.

Where to see Mountains, open woodlands, hedgerows, moorlands and heaths; visits open, scrubby areas in winter. Common resident.

House Sparrow *Passer domesticus* 15cm

Male light grey on underside, streaky brown above, with black chin and bib, dark lores, grey crown bordered with brown. Female lighter brown, lacks black head markings, has pale supercilium. Calls are various loud chirps and chatters. Very social, foraging and resting in groups, often with other seed-eating birds, and nesting in loose colonies. Frequently dust-bathes.

Where to see Urban environments and around farms; visits grain stores and harvested crop fields. In Italy resident only in far north; elsewhere replaced by Italian Sparrow.

Italian Sparrow *Passer italiae* 15cm

Very closely related to House Sparrow; females of the two species are indistinguishable. Male Italian Sparrow like male House Sparrow but has entirely chestnut-brown crown and whiter cheeks; black of lores extends behind eye as short eye-stripe. Underside a slightly paler shade of grey. Voice, behaviour and biology as that of House Sparrow. The Italian Sparrow constitutes a stable hybrid population between House Sparrow and Spanish Sparrow (the latter a common bird in southern Europe, both east and west of Italy), and is treated as a full species.

Where to see A common resident in towns and open countryside in mainland Italy and Sicily.

♀

♂ non-br.

♂ br.

Spanish Sparrow *Passer hispaniolensis* 15cm

Closely related to House and Italian Sparrows; females usually indistinguishable though may show hint of dark grey streaking on flanks. Male distinctive and very boldly patterned in black, chestnut and whitish, in summer particularly almost lacks dull grey tones of other species. Crown chestnut, narrow white supercilium and black eye-stripe. Black bib spreads as heavy streaking down breast, onto flanks and sides of undertail, also on back. Wings chestnut with double white wing-bar. In winter pattern more subdued. Voice and habits similar to House and Italian Sparrows.

Where to see Prefers more rural habitats. Pure birds are common resident on Sardinia, uncommon and local in south-east mainland Italy and Po delta.

♂

♀

♂

Tree Sparrow *Passer montanus* 13cm

The smallest sparrow; sexes identical. Boldly and neatly marked. Crown chestnut, neck-sides and cheek white, with black spot below eye. Small black bib. Underside light grey-brown, upperside streaky chestnut with narrow double white wing-bar. Juvenile more muted but has same face pattern. Has chirping calls, also distinctive bright two-note *ts-wit*. Feeds on insects and seeds; gregarious. Will flock with other small seed-eaters.

Where to see Farmland, villages and small towns. Common and widespread resident, including on Sardinia and Sicily.

Rock Sparrow *Petronia petronia* 15cm

Recalls a stocky female House Sparrow, but with stronger markings and streaky flanks and breast-sides. Has bold dark eye-stripe and cheek outline, dark edges to crown, and contrasting pale fringes to dark wing feathers. Bill dark grey with yellow base, legs pinkish. Gives various chirping calls, similar to those of other sparrows, also distinctive, nasal and more finch-like *dvee-zu*. Will mix with other sparrow species.

Where to see Rugged, rocky upland habitats, sometimes around habitation. Uncommon and localised resident in southern Italy, Sardinia and Sicily.

White-winged Snowfinch *Montifringilla nivalis* 17cm

non-br.

A relatively large and distinctive sparrow-like bird of mountainous regions. Rather slim and long-bodied, with long tail. White on underparts, tail-sides and most of wings; head grey, back grey-brown with faint darker streaks. In breeding plumage has black bib and black bill (dull yellow in winter). In flight shows bold pattern of white, black-tipped wings, and black, white-edged tail. Has varied calls including mewing, chattering and rolling notes; song a jerky twittering. Often forages around ski lodges and can be very approachable.

Where to see A bird of high altitudes, rarely seen below 1,500m. Resident in Alps and Apennines.

Grey Wagtail *Motacilla cinerea* 19cm

♂ br.

Largish with very long tail. Upperparts plain blue-grey, underparts yellow (brightest under tail). Lacks white wing-bars but tertials white-fringed. Narrow white supercilium and white outer tail feathers; breeding male has black throat. Juvenile has yellow restricted to undertail; breast pale with pinkish tint. Legs greyish-pink. Call a sharp two-note *chi-tick*; song short phrase of hissing notes. Very active, constantly bobbing tail, catches flies over water.

Where to see Fast-flowing, rocky streams, rivers; in winter also on lake shores, sometimes in towns. Common, widespread resident.

White Wagtail *Motacilla alba* 17cm

br.

non-br.

Dapper monochrome wagtail. Back grey, underparts white, wings grey with darker feather centres and double white wing-bar. In breeding plumage, head white with black hind-crown and neck, large black bib; male more crisply marked than female. In winter, throat white, crown grey, bib reduced to black breast-band. Juvenile duskier with yellow tint on face. Legs black. Call a shrill *chissick*, song a simple twitter. Has sprightly gait, tail constantly bobbing.

Where to see Open countryside, often near water, urban areas. Common, widespread resident.

br.

Yellow Wagtail *Motacilla flava cinereocapilla/M. f. feldegg* 15cm

♂ *cinereocapilla*

♂ *feldegg*

Many subspecies occur in Europe; *cinereocapilla* (Ashy-headed Wagtail) breeds commonly in Italy. Male has dark grey head, blackish cheeks and white chin, otherwise olive-green above and bright yellow below. Female muted, with light grey head and pale supercilium. Has white outer tail-feathers and narrow double white wing-bars; legs black. Subspecies *feldegg* (Black-headed Wagtail) is scarce, localised breeder, mainly in south. Call high *tsit*, song short phrase of dry notes.

Where to see Damp lowland grassland and marshes. Fairly widespread summer visitor and passage migrant.

Tawny Pipit *Anthus campestris* 17cm

Large, pale, long-legged wagtail-like pipit. Pale grey-brown fading to whitish on breast and belly. Wing feathers pale with dark centres. Faintly streaked on back and breast-sides. Face well marked with dark eye-stripe, pale supercilium. Juvenile darker, with dark scalloping on back and streaks on breast. Call a sparrow-like *chirrup*, song a repeated two- or three-note phrase. Has strutting walk, searching on ground for insect prey.

Where to see Sparsely vegetated dry, open, flat ground in highlands and lowlands. Fairly common summer visitor.

Meadow Pipit *Anthus pratensis* 15cm

(Near Threatened) Very like Tree Pipit. Has less well-marked face. Streaking on flanks is same thickness as on breast. Breast streaking sometimes coalesces to form solid dark patch. Legs pinkish; strikingly elongated hindclaw. Call high *weet weet weet*, given when taking flight. Song similar notes, often given in rising song-flight that begins and ends at ground level. Forages mostly on ground, walking or running in pursuit of insect prey; somewhat gregarious.

Where to see Grassland, farmland, moors, heaths. Widespread winter visitor to Italy.

Tree Pipit *Anthus trivialis* 15cm

Pale, neatly patterned. Upperparts light brown with black streaks, underparts whitish-buff with black streaks on breast, becoming much finer on flanks. Slight dark eye-stripe and pale supercilium. Legs pinkish, hindclaw shorter than in Meadow Pipit. Bill somewhat stout for a pipit, with pink base. Song a long phrase of varied trills, often given in parachuting song-flight from treetop to ground level. Call slightly hoarse *spizzz*.

Where to see Mountain woodland edges, open country with some tall trees. Fairly widespread summer visitor and passage migrant.

Red-throated Pipit *Anthus cervinus* 15cm

br.

non-br.

Adult in breeding plumage has variably extensive brick-red throat and face but in other plumages is very similar to Meadow and Tree Pipits. Plumage somewhat more strongly contrasting, with paler fringes to wing feathers and usually heavier streaking on breast and flanks than Meadow Pipit. Hindclaw not noticeably long. Best identified by call, a drawn-out, single *speeeeeeh* that drops in pitch at the end.

Where to see Grassland and other open country, especially near coast. Uncommon passage migrant.

Water Pipit *Anthus spinoletta* 16cm

br.

non-br.

Larger than Meadow Pipit. In breeding plumage rather plain grey on crown, cheeks and back; wings dark with bold pale feather fringes. Has strong white supercilium. Underparts whitish, with faint darker streaking on flanks, variable pink flush to chin and breast. Legs dark. In winter duller with strong streaks on underside. Song protracted, varied trilling, call sharp *pssit*.

Where to see Breeds on montane slopes and fields; moves to marshy lowlands in winter. Localised summer visitor to mountains; much more widespread in winter.

Chaffinch *Fringilla coelebs* 15cm

Sparrow-sized. Male colourful, with pink cheeks and underparts, maroon back, blue-grey crown, black wings with white shoulder and white wing-bar that forms a 'T' shape. Rump greenish, tail dark with white outer feathers. Generally duller in winter. Female has similar wing and tail pattern, but body plumage drab grey-brown with no strong facial markings. Call bright *twink* or more mournful *dwee*, song fast, descending chirruped phrase. Feeds on insects in summer, seeds in winter.

Where to see Open woodlands, gardens and parkland. Common, widespread resident.

Brambling *Fringilla montifringilla* 15cm

Like Chaffinch in shape and pattern, but breast orange, shading to white on belly with black spots on flanks; shoulder patch orange. Head grey with dark crown and neck-sides; male also has blackish markings around eye which become more extensive as spring approaches. Black tail, white rump. Call harsh two-note *te-chup*. Sociable and often joins flocks of Chaffinches; feeds on seeds, especially beech mast.

Where to see Winter visitor; numbers vary year on year (sometimes abundant). More likely in north and in mountains.

Hawfinch *Coccothraustes coccothraustes* 17cm

Big, stocky finch. Short tail and very large head and bill give unique outline. Mainly pinkish-orange, with dark back, whitish shoulder patch and blue-grey on flight feathers. Head orange, neck grey, has small black bib and black lores. Bill blue-grey. Tail has broad white tip. Female slightly drabber than male. Call a hard, buzzing or grating tsick, song a quiet series of similar notes. A shy bird, tending to keep to the high treetops. Fond of cherries and able to crack their stones.

Where to see Found in undisturbed mixed woodland, particularly with some fruit trees. Patchily distributed resident in hills and mountains, in winter may visit grasslands and towns.

Bullfinch *Pyrrhula pyrrhula* 16cm

Stocky with short but very stout bill. Male has bright pink cheeks and underparts, black cap, grey back, black wings with white wing-bar, white rump, black tail. Female similar but muted; underparts soft pinkish-grey. Juvenile much browner, lacks black head markings. Call soft, hesitant single whistle. Song quiet, combines fluting notes with squeaks and rattles. Usually seen in pairs or small groups, shy and inconspicuous.

Where to see Woodland, parks, larger gardens in Alps and Apennines. Fairly widespread resident, wandering more widely in winter.

Greenfinch *Chloris chloris* 15cm

Stout, large-billed and large-headed finch. Male quite uniform mossy green, with hints of grey on cheeks, flanks and wings; some look very yellow. Has dark flight feathers and yellow along wing edge and tail-sides. Bill rather pale. Female similar but greyer; juvenile strongly streaked on both upperparts and underparts. Call a rather metallic chirping note. Song varied, with fast twitters and drawn-out downslurred notes, sometimes given in circling, butterfly-like song-flight. Feeds on ground and in trees. Often gregarious in winter, may be aggressive to other finches.

Where to see Found in woodland, hedgerows, gardens and parks. A common, widespread resident.

Linnet *Linaria cannabina* 13cm

Smallish, long-tailed. Male has grey head with red forehead patch, chestnut-brown back, pale underparts with pink breast, brightest in summer. Female grey-brown with light streaks on back and breast; juvenile more heavily streaked. Bill and legs grey. All plumages have white panel along wing edge and pale cheek patch.

Call bright, buzzy *tit-it*, song series of twittered phrases. Forms large flocks in winter, which visit weedy fields to eat seeds.

Where to see Farmland, meadows with hedgerows. A fairly common and widespread resident.

Lesser Redpoll *Acanthis cabaret* 12cm

Small, agile, fork-tailed. Plumage light grey-brown with darker streaks on upperparts and underparts. Has broad pale wing-bar. Small red forehead patch, black lores and small black bib, short black eye-stripe. Breast flushed reddish-pink in adult males. Bill yellow with black tip. Call hard *tet tet*. Feeds mainly in high treetops, extracting seeds from alder cones or similar, often dangling upside-down. Usually in flocks, sometimes with Siskins.

Where to see Woodland, parks and gardens. Rather common resident in Alps; some visit northern plains in winter.

Common Crossbill *Loxia curvirostra* 16cm

Robust, short-tailed finch, bill mandible tips elongated and crossing over. Male rich red, almost unmarked, with darker wings and tail. Female mossy green, with yellowish rump. Juvenile rather grey, and heavily streaked all over. Call distinctive metallic *klip*. A specialist feeder on pine cones, which it grips in its foot while prising the scales apart with its bill. Usually in small flocks.

Where to see Coniferous forest or smaller stands of pines in more open countryside. Localised resident but has periodic irruptive movements.

Goldfinch *Carduelis carduelis* 13cm

Distinctive, colourful, slim finch. Body light brown, shading to white on belly. Face white with red around bill base, black crown and cheek-sides. Wings black with yellow wing-bar and white spots in wingtip; rump white, tail black with white spots at tip. Juvenile streaked light brown with adult-like wings and tail. Bill quite slim, pale with dark tip. Call laughing twitter, song series of similar notes. Semi-colonial nester, gregarious in autumn and winter.

Where to see Woodland edges, hedgerows, weedy meadows, parks, gardens. Common, widespread resident.

Citril Finch *Carduelis citrinella* 12cm

♂

Dainty small finch with short bill, very like closely related Corsican Finch. Male predominantly yellow on underparts, greener on upperparts. Double yellow wing-bar. Dark eye-stripe, light grey hind-neck and cheek-sides. Female has more extensive grey, suffusing back, breast-sides, flanks and undertail; back has dark streaking and wing-bars narrower. Call, given in flight, high *cheeet*. Feeds mainly in trees, eating seeds and some insects.

Where to see Upland coniferous forests and adjacent meadows. Resident in the Alps.

Serin *Serinus serinus* 11cm

♀

♂

Tiny, short-billed finch. Male has bright yellow face and breast, whiter on belly; flanks have bold dark streaks. Upperparts green with black streaks on flanks and breast-sides, wings and tail blackish, showing yellow rump and tail-sides in flight. Female similar but duller yellow-green. Juvenile browner, heavily streaked. Call a high-pitched buzzing trill; song similar with shivering, electrical quality. Feeds on ground and in treetops. Usually alone or in pairs.

Where to see Light woodland, parks, gardens. Common and widespread resident; summer visitor in far north.

Corsican Finch *Carduelis corsicana* 12cm

Small slim finch. Front of face, rump and entire underparts bright yellow, back of neck grey, back brown with darker streaks, wings black with broad, double wing-bars and feather edges yellow-green. Bill dark. Female drabber than male. Calls are short *te* or *teh* notes; song a series of trills of varying speed. A close relative of more northerly Citril Finch and formerly considered a subspecies of it. Tends to forage on ground. Feeds on seeds and insects.

Where to see Occurs only on Corsica, Sardinia, Elba, Capraia and Gorgona Islands. Found in scrubby heathland and forest edges, from sea level to the mountains; resident.

Siskin *Spinus spinus* 12cm

Small with rather long, pointed bill. Male boldly marked with yellow face, breast, wing-bars, tail-sides and rump; crown, bib and most of wings blackish, back green with fine dark streaks; belly white with dark streaks on flanks. Female has less yellow and lacks black face markings. Juvenile greyer, heavily streaked. Call nasal, two-note *dzwee* or *dzee-oo*. Song series of varied, high-pitched trills. In winter travels in flocks and feeds in high treetops.

Where to see Coniferous or mixed woodland. Localised resident, widespread in winter.

Snow Bunting *Plectrophenax nivalis* 16cm

♀ non-br.

Has extensive white in plumage, and very long wings. Non-breeding birds have sandy-brown markings on head and breast-sides, back light brown with darker feather centres, primaries and tail centre black; underparts white. Male whiter than female, which may have light brown wash over face, breast and flanks. Shows white on inner part of wing upperside in flight. Call a tuneful rippling note. Feeds in flocks on ground, very approachable.

Where to see Sea coasts, sometimes inland in open rocky areas with short grass. Rare passage migrant and winter visitor. Can flock with White-winged Snowfinch in the Alps.

♂ non-br.

Black-headed Bunting *Emberiza melanocephala* 16cm

A large, colourful bunting. Male has black head, bright yellow chin, collar and underside, and chestnut back. Wings dark with broad pale feather fringes and tips, creating double white wing-bars. Female drab with grey head shading to grey-brown back, no collar. Juvenile much greyer, with just a little yellow on throat and belly. Has various brief, dry calls; song is simple dry, rattling trill, given from top of bush or wire. Often forages on ground.

Where to see A bird of open countryside with scrub, scattered trees or hedgerows. Summer visitor to south-east Italy, a few small colonies elsewhere; not Sicily or Sardinia.

♂ imm.

♀

♂ br.

Corn Bunting *Emberiza calandra* 18cm

Large, stocky, long-tailed bunting with very plain plumage. Light earth-brown on upperparts, a shade paler below, marked with broad darker streaks on crown, back and shoulders; fine streaks on underparts. Face rather plain but with pale stripe outlining cheek, fine dark stripe below. Bill stout, grey at tip, pinkish at base. Often dangles legs in flight. Call is hard, metallic *tsriiit*, song a short rattle likened to a bunch of keys being shaken. Nests at ground level; male uses bush, fencepost or overhead wire as song perch.

Where to see Open countryside, particularly farmland. Common, widespread resident.

Rock Bunting *Emberiza cia* 16cm

non-br.

Slim, medium-sized bunting. Grey head and breast, rest of plumage reddish-brown. Head well marked with black eye-stripe, also black stripes on crown-sides and outlining cheek. Back has heavy dark streaking, wing feathers dark-centred and pale-tipped to form double white wing-bar. Underside unmarked. Female drabber than male. Call a sharp *tsi* or more drawn-out *tsiiu*, song sweet, clear and melodic. Can be quite tame and confiding, though unobtrusive as it feeds quietly on ground.

Where to see Open, rugged and rocky countryside with sparse scrub. Resident in montane regions, more widespread in winter.

Cirl Bunting *Emberiza cirlus* 16cm

Male colourful with yellow on face and yellow belly, plus dark green crown, collar and breast-band, and black throat and eye-stripe. Flanks streaked rusty-brown. Upperparts greenish-brown, with streaked back and wings. Female much drabber, lacking black head markings, more streaked on underparts; best told from female Yellowhammer by grey-green rather than chestnut-brown rump. Juvenile whitish on belly with heavy streaking. Call a hard *zit*; song a simple dry rattle recalling Lesser Whitethroat.

Where to see Farmland with hedgerows, vineyards, large gardens. Common, widespread resident.

Yellowhammer *Emberiza citrinella* 16cm

Male has yellow head and underparts, with narrow dark eye-stripe, cheek-stripe and crown streaks (some have solid yellow head). Breast-sides rusty chestnut, extending in streaks along flanks. Upperparts light grey-brown, with darker streaks. Rump reddish-brown. Female and juvenile much drabber and streakier, though still with yellowish overall look. Has dry, brief calls. Song a short rattle with drawn-out final note.

Where to see Farmland with hedgerows, scrubby open countryside. Common resident (scarce in south), more widespread in winter. Not on Sardinia.

Ortolan Bunting *Emberiza hortulana* 16cm

Slim, long-tailed bunting. Head and breast soft grey-green, with yellowish throat and cheek-stripe; yellow eye-ring strongly accentuates dark eye. Underparts unmarked orange-brown. Upperparts grey-brown; wing feathers have dark centres and pale fringes. Juvenile spotted on breast and flanks. Bill and legs pinkish. Has various brief, metallic calls, and clear but simple ringing song of evenly spaced notes.

Where to see Woodland edges, farmland with patches of woodland, hillsides and mountains. Scarce and declining summer visitor mainly to northern and central Italy, passage migrant in south.

Reed Bunting *Emberiza schoeniclus* 14cm

♀

♂ br.

Smallish, sparrow-like. Male in breeding plumage has black head and breast, with white collar and cheek-stripe; almost unmarked pale underparts. Upperparts dark brown, streaked with black. Outer tail-feathers white. Female has brown rather than black head, with pale chin and supercilium, grey rather than white collar, underside strongly streaked. Short, buzzy calls; song comprises two distinct notes followed by a short jangling phrase.

Where to see Wetland areas with reedbeds, may move to farmland in winter. Rather localised resident, more widespread in winter.

GLOSSARY AND FURTHER READING

Glossary

Bare parts The unfeathered parts of a bird – legs, bill and sometimes parts of the face.

Bib A patch of colour covering the throat and upper breast.

Breast-band A stripe of colour across the breast.

Call A simple sound made by a bird, for contact or to warn of danger.

Crest A tuft of feathers on top of a bird's head.

Drumming (in woodpeckers) Repeatedly striking a resonant tree branch or trunk with the bill; equivalent to song.

Ear-tufts Tufts of feathers on either side of the top of a bird's head.

Eye-mask A patch of contrasting colour over the eye and upper cheek.

Eye-ring A circle of coloured feathers or bare skin around the eye.

Eye-stripe A line (usually dark) of colour running in front of and behind the eye.

Facial disc The face of an owl, outlined by a ruff of short, stiff feathers.

Feral Describes wild-living birds that are descended from captive stock.

Fringes The edges of feathers, often contrastingly coloured to give a scaly appearance.

Invertebrate Animals without a backbone – insects, spiders, snails etc.

Iridescent Having a brightly coloured sheen in certain lights.

Leading edge (of wing) The front edge of the opened wing as the bird is flying.

Lores Area between the base of the bill and the eye.

Moustachial stripe A (usually dark) stripe running down the cheek.

Passage migrant A bird that occurs in a certain area only during its migratory journey – passing through.

Passerine Songbird or 'perching bird'.

Primaries The outermost flight feathers, or 'hand' of the wing.

Resident A bird that is present in a certain area year-round.

Scrub Any habitat with plenty of bushes but few or no trees.

Secondaries The inner flight feathers, or 'arm' of the wing.

Song Sound made by a bird (usually male) to advertise its territory; usually more complex than calls.

Song-flight A particular type of flight (usually rising then falling) performed as the bird sings.

Subsong Variation of typical song, usually quieter.

Supercilium A line of colour (usually pale) above the eye, in an 'eyebrow' position.

Tail-band A stripe of colour across the tail, most often the tail-tip.

Trailing edge (of wing) The rear edge of the opened wing as a bird flies.

Upperparts The top side of a bird – usually the crown, neck, back, upperside of wings, rump and upperside of tail.

Underparts The bottom side of a bird – usually throat, breast, belly and flanks.

Wattle A conspicuous fleshy growth of coloured skin on a bird's face.

Wing-bar A stripe of contrasting colour on the wing, usually visible when wing is folded as well as spread.

Further reading

The following books should be of interest to those wishing to learn more about the birds, birdwatching and other wildlife of Italy:

Jepson, Tim, *Wild Italy: A Traveller's Guide*, Sheldrake Press, London, 2005. A detailed site guide for places to enjoy wildlife-watching, hiking and other outdoor activities in Italy.

Lega Italiana Protezione Uccelli, *Where to Watch Birds in Italy*, Christopher Helm, London, 1994. A guide to the best birdwatching sites in Italy.

Price, Gillian. *Walking in Sicily*, Cicerone Guides, Kendal, 2014. One of several walking guides for Italy published by Cicerone Press, giving tried-and-tested walks of varied length and difficulty through beautiful wild scenery.

Svensson, Lars, *Collins Bird Guide*, Collins, London, 2022 (3rd ed). Detailed and comprehensive identification guide to birds of Britain and Europe, including most species recorded in Italy to date.

PHOTO CREDITS AND ACKNOWLEDGEMENTS

Photo credits

All the photographs in this book were taken by Daniele Occhiato, except for those listed below.

Bloomsbury Publishing would like to thank the following for providing photographs and for permission to reproduce copyright material within this book. While every effort has been made to trace and acknowledge all copyright holders, we would like to apologise for any errors and omissions, and invite readers to inform us so that corrections can be made to future editions.

Key: T = top; C = centre; B = bottom; L = left; R = right; BL = bottom left; BR = bottom right; TL = top left; TR = top right; CL = centre left; CR = centre right.

8 afinocchiaro/iStock; **11** only_fabrizio/iStock; **12** antonioscarpi/iStock; **13** AdventureSte/iStock; **16** Isaac74/iStock; **21** T Enrico Benussi; **25** TR Piotr Krzeslak/Shutterstock; **30** BL Natural Imaging/Shutterstock; **31** BL rock ptarmigan/Shutterstock; BR Bouke Atema/Shutterstock; **32** BL Susan Hodgson/Shutterstock; **33** TR Stephen Catterall/iStock; **42** BR Erni/Shutterstock; **45** T Sylvia Adams/Shutterstock; **47** B Gelpi/iStock; **78** TL Agami Photo Agency/Shutterstock; TR Agami Photo Agency/Shutterstock; **86** TL Carlos Naza Bocos; **93** TL birdsonline/iStock; TR Rafael Armada/Agami; BR David Havel/Shutterstock; **105** B AOosthuizen/iStock; **108** BR Dr Pankaj Maheria/Shutterstock; **109** T Santiago Urquijo/Shutterstock; B Michele Mendi; **111** T usluomer/Shutterstock; C Marc Guyt/iStock; **114** T Saverio Gatto; **117** B Enrico Benussi; **122** BL imageBROKER/Friedhelm Adam/Getty; BR Andreas Rose/Shutterstock; **127** B Enrico Benussi; **128** T Enrico Benussi; B Enrico Benussi; **130** BL Enrico Benussi; BR Enrico Benussi; **131** T Michele Mendi; **135** T Fabrizio Moglia/Getty; B Agami Photo Agency/Shutterstock; **139** B Marco Sbrò/500px/Getty; **140** B WildMedia/Shutterstock; **144** T Michele Mendi; **147** T Michele Mendi; **150** T Lorenzo Magnolfi; **188** TL Paolino Massimiliano Manuel/iStock; TR Oscar Díez/Agami; **212** TL Michele Mendi; TR Michele Mendi; **213** T Joan Padro/Shutterstock.

Acknowledgements

Marianne Taylor would like to thank Jim Martin, Jane Lawes, Jenny Campbell and Amy Hodkin at Bloomsbury for guiding this book from conception to completion. She would also like to thank Daniele Occhiato for his superb photographs and invaluable feedback on the text, and Rod Teasdale, Tim Harris, Guy Kirwan and Angie Hipkin for their work on the layout, editing, proofreading and indexing respectively.

INDEX

Acanthis cabaret 211
Accentor, Alpine 198
Accipiter gentilis 112
nisus 113
Acrocephalus arundinaceus 162
melanopogon 160
palustris 161
schoenobaenus 160
scirpaceus 161
Actitis hypoleucos 73
Aegithalos caudatus 170
Aegolius funereus 122
Alauda arvensis 156
Alca torda 79
Alcedo atthis 124
Alectoris barbara 39
graeca 40
rufa 40
Amandava amandava 198
Anas acuta 27
crecca 27
platyrhynchos 26
Anser albifrons 21
anser 20
serrirostris 21
Anthus campestris 205
cervinus 207
pratensis 206
spinoletta 207
trivialis 206
Apus apus 51
pallidus 51
Aquila chrysaetos 109
fasciata 108
Ardea alba 99
cinerea 98
purpurea 98
Ardeola ralloides 101
Arenaria interpres 66
Asio flammeus 122
otus 121
Athene noctua 120
Avadavat, Red 198
Avocet 59
Aythya ferina 28
fuligula 30
marila 30
nyroca 29

Bee-eater, European 125
Bittern 97
Little 97
Blackbird 187
Blackcap 171
Bluethroat 190
Botaurus stellaris 97
Brambling 208
Bubo bubo 117
Bubulcus ibis 100
Bucephala clangula 32
Bullfinch 209
Bunting, Black-headed 216
Cirl 218
Corn 217
Ortolan 219
Reed 219
Rock 217
Snow 215
Burhinus oedicnemus 57
Bustard, Little 47
Buteo buteo 116
rufinus 117
Buzzard, Common 116
Long-legged 117

Calandrella brachydactyla 153
Calidris alba 69
alpina 69
canutus 67
ferruginea 68
minuta 70
pugnax 67
temminckii 68
Calonectris diomedea 91
Capercaillie 35
Caprimulgus europaeus 48
Carduelis carduelis 212
citrinella 213
corsicana 214
Cecropis daurica 166
Certhia brachydactyla 182
familiaris 181
Cettia cetti 171
Chaffinch 208
Charadrius alexandrinus 62
dubius 63
hiaticula 63
morinellus 64
Chiffchaff 168
Chlidonias hybrida 88
leucopterus 88
niger 87
Chloris chloris 210
Chough, Alpine 145
Red-billed 145
Chroicocephalus genei 80
ridibundus 81
Ciconia ciconia 94
nigra 93s
Cinclus cinclus 183
Circaetus gallicus 107
Circus aeruginosus 110
cyaneus 110
macrourus 111
pygargus 112
Cisticola juncidis 158
Cisticola, Zitting 158
Clamator glandarius 48
Clanga clanga 107
Coccothraustes coccothraustes 209
Columba livia 44
oenas 45
palumbus 45
Coot 54
Coracias garrulus 126
Cormorant 96
Pygmy 95
Corncrake 52
Corvus corax 148
cornix 147
corone 147
frugilegus 146
monedula 146
Coturnix coturnix 38
Crake, Little 55
Spotted 53
Crane, Common 56
Crex crex 52
Crossbill, Common 212
Crow, Carrion 147
Hooded 147
Cuckoo, Common 49
Great Spotted 48
Cuculus canorus 49
Curlew 65
Curruca cantillans 174
communis 176
conspicillata 177
curruca 173
hortensis 173
melanocephala 174
nisoria 172
sarda 176
subalpina 175
undata 178
Cyanecula svecica 190
Cyanistes caeruleus 151
Cygnus cygnus 22
olor 22

Delichon urbicum 166
Dendrocopos leucotos 128
major 129
medius 128
Dipper 183
Diver, Black-throated 90
Red-throated 90
Dotterel 64
Dove, Collared 47
Rock 44
Stock 45
Turtle 46
Dryobates minor 130
Dryocopus martius 131
Duck, Ferruginous 29
Tufted 30
Dunlin 69
Dunnock 199

Eagle, Bonelli's 108
Booted 108
Golden 109
Greater Spotted 107
Short-toed 107
White-tailed 116
Egret, Cattle 100
Great White 99
Little 100
Egretta garzetta 100
Eider, Common 31
Emberiza calandra 217
cia 217
cirlus 218
citrinella 218
hortulana 219
melanocephala 216
schoeniclus 219
Erithacus rubecula 189

Falco biarmicus 136
cherrug 138
columbarius 136
eleonorae 135
naumanni 133
peregrinus 138
subbuteo 137
tinnunculus 132
vespertinus 134
Falcon, Eleonora's 135
Lanner 136
Peregrine 138
Red-footed 134

Saker 138
Ficedula albicollis 190
hypoleuca 191
Fieldfare 187
Finch, Citril 213
Corsican 214
Firecrest 180
Flamingo, Greater 41
Flycatcher, Collared 190
Pied 191
Spotted 188
Thyrrenian 188
Fringilla coelebs 208
montifringilla 208
Fulica atra 54

Gadwall 25
Galerida cristata 156
Gallinago gallinago 72
media 72
Gallinula chloropus 54
Gannet, Northern 95
Garganey 24
Garrulus glandarius 143
Gavia arctica 90
stellata 90
Gelochelidon nilotica 87
Glareola pratincola 77
Glaucidium passerinum 119
Godwit, Bar-tailed 65
Black-tailed 66
Goldcrest 179
Goldeneye 32
Goldfinch 212
Goosander 33
Goose, Greater White-fronted 21
Greylag 20
Tundra Bean 21
Goshawk 112
Grebe, Black-necked 44
Great Crested 43
Little 42
Red-necked 43
Slavonian 42
Greenfinch 210
Greenshank 75
Grouse, Black 36
Hazel 34
Grus grus 56
Gull, Audouin's 83
Black-headed 81
Caspian 85
Common 82
Herring 84
Lesser Black-backed 85
Little 81
Mediterranean 82
Slender-billed 80
Yellow-legged 84
Gypaetus barbatus 105
Gyps fulvus 106

Haematopus ostralegus 60
Haliaeetus albicilla 116
Harrier, Hen 110
Marsh 110
Montagu's 112
Pallid 111
Hawfinch 209
Heron, Grey 98
Night 101
Purple 98
Squacco 101
Hieraaetus pennatus 108
Himantopus himantopus 58
Hippolais icterina 158
polyglotta 159
Hirundo rustica 165
Hobby 137
Honey-buzzard, European 104
Hoopoe 123
Hydrobates pelagicus 93
Hydrocoleus minutus 81
Hydroprogne caspia 86

Ibis, African Sacred 102
Glossy 103
Ichthyaetus audouinii 83
melanocephalus 82
Ixobrychus minutus 97

Jackdaw 146
Jay 143
Jynx torquilla 127

Kestrel, Common 132
Lesser 133
Kingfisher 124
Kite, Black 115
Red 114
Kittiwake 79
Knot 67

Lagopus muta 37
Lanius collurio 141
excubitor 141
minor 142
senator 142
Lapwing 61
Lark, Calandra 155
Crested 156
Short-toed 153
Larus argentatus 84
cacchinans 85
canus 82
fuscus 85
michahellis 84
Leiothrix lutea 179
Leiothrix, Red-billed 179
Limosa lapponica 65
limosa 66
Linaria cannabina 211
Linnet 211
Locustella luscinioides 163
naevia 163
Lophophanes cristatus 149
Loxia curvirostra 212
Lullula arborea 154
Luscinia megarhynchos 189
Lymnocryptes minimus 71
Lyrurus tetrix 36

Magpie 143
Mallard 26
Mareca penelope 25
strepera 25
Martin, Crag 164
House 166
Sand 164
Melanitta fusca 31
nigra 32
Melanocorypha calandra 155
Merganser, Red-breasted 33
Mergus merganser 33
serrator 33
Merlin 136
Merops apiaster 125
Microcarbo pygmaeus 95
Milvus migrans 115
milvus 114
Monticola saxatilis 193
solitarius 194
Montifringilla nivalis 203
Moorhen 54
Morus bassanus 95
Motacilla alba 204
cinerea 203
flava 205
Muscicapa striata 188
Myiopsitta monachus 139

Neophron percnopterus 105
Netta rufina 28
Nightingale 189
Nightjar, European 48
Nucifraga caryocatactes 144
Numenius arquata 65
phaeopus 64
Nutcracker 144
Nuthatch 181
Nycticorax nycticorax 101

Oenanthe isabellina 196
melanoleuca 197
oenanthe 196
Oriole, Golden 140
Oriolus oriolus 140
Osprey 104
Otus scops 119
Owl, Barn 118
Eagle 117
Eurasian Scops 119
Little 120
Long-eared 121
Pygmy 119
Short-eared 122
Tawny 121
Tengmalm's 122
Ural 120
Oystercatcher 60

Pandion haliaetus 104
Panurus biarmicus 157
Parakeet, Monk 139
Rose-ringed 139
Partridge, Barbary 39
Grey 36
Red-legged 40
Rock 40
Parus major 153
Passer domesticus 199
hispaniolensis 201
italiae 200
montanus 202
Pastor roseus 184
Perdix perdix 36
Periparus ater 149
Pernis apivorus 104
Petronia petronia 202
Phalacrocorax aristotelis 96
carbo 96
Phalarope, Red-necked 73
Phalaropus lobatus 73
Phasianus colchicus 38
Pheasant, Common 38
Phoenicopterus roseus 41
Phoenicurus ochruros 192
phoenicurus 193
Phylloscopus bonelli 169
collybita 168
inornatus 167
sibilatrix 167
trochilus 169
Pica pica 143
Picoides tridactylus 127
Picus canus 130
viridis 132
Pigeon, Feral 44
Pintail 27
Pipit, Meadow 206
Red-throated 207
Tawny 205
Tree 206
Water 207
Platalea leucorodia 103
Plectrophenax nivalis 215
Plegadis falcinellus 103

Plover, Golden 61
Grey 60
Kentish 62
Little Ringed 63
Ringed 63
Pluvialis apricaria 61
squatarola 60
Pochard 28
Red-crested 28
Podiceps auritus 42
cristatus 43
grisegena 43
nigricollis 44
Poecile montanus 151
palustris 150
Porphyrio porphyrio 55
Porzana porzana 53
Pratincole, Collared 77
Prunella collaris 198
modularis 199
Psittacula krameri 139
Ptarmigan 37
Ptyonoprogne rupestris 164
Puffinus yelkouan 92
Pyrrhocorax graculus 145
pyrrhocorax 145
Pyrrhula pyrrhula 209

Quail, Common 38

Rail, Water 52
Rallus aquaticus 52
Raven 148
Razorbill 79
Recurvirostra avosetta 59
Redpoll, Lesser 211
Redshank, Common 76
Spotted 74
Redstart, Black 192
Common 193
Redwing 186
Reedling, Bearded 157
Regulus ignicapilla 180
regulus 179
Remiz pendulinus 152
Ring Ouzel 188
Riparia riparia 164
Rissa tridactyla 79
Robin 189
Roller 126
Rook 146
Ruff 67

Sanderling 69
Sandpiper, Common 73
Curlew 68
Green 74
Marsh 75
Wood 76
Saxicola rubetra 195
rubicola 194
Scaup 30
Scolopax rusticola 71
Scoter, Common 32
Velvet 31
Serin 213
Serinus serinus 213
Shag 96
Shearwater, Scopoli's 91
Yelkouan 92
Shelduck 23
Ruddy 23
Shoveler 24
Shrike, Great Grey 141
Lesser Grey 142
Red-backed 141
Woodchat 142
Siskin 214
Sitta europaea 181
Skua, Arctic 78
Pomarine 78
Skylark 156
Snipe, Common 72
Great 72
Jack 71
Snowfinch, White-winged 203
Somateria mollissima 31
Sparrow, House 199
Italian 200
Rock 202
Spanish 201
Tree 202
Sparrowhawk 113
Spatula clypeata 24
querquedula 24
Spinus spinus 214
Spoonbill, Eurasian 103
Starling 183
Rose-coloured 184
Spotless 185
Stercorarius parasiticus 78
pomarinus 78
Sterna hirundo 89
Sternula albifrons 86
Stilt, Black-winged 58
Stint, Little 70
Temminck's 68
Stonechat 194
Stone-curlew 57
Stork, Black 93
White 94
Storm-petrel, European 93
Streptopelia decaocto 47
turtur 46
Strix aluco 121
uralensis 120
Sturnus unicolor 185
vulgaris 183
Swallow, Barn 165
Red-rumped 166
Swamphen, Purple 55
Swan, Mute 22
Whooper 22
Swift, Alpine 50
Common 51
Pallid 51
Sylvia atricapilla 171
borin 172

Tachybaptus ruficollis 42
Tachymarptis melba 50
Tadorna ferruginea 23
tadorna 23
Teal, Eurasian 27
Tern, Black 87
Caspian 86
Common 89
Gull-billed 87
Little 86
Sandwich 89
Whiskered 88
White-winged Black 88
Tetrao urogallus 35
Tetrastes bonasia 34
Tetrax tetrax 47
Thalasseus sandvicensis 89
Threskiornis aethiopicus 102
Thrush, Blue Rock 194
Common Rock 193
Mistle 185
Song 186
Tichodroma muraria 180
Tit, Blue 151
Coal 149
Crested 149
Great 153
Long-tailed 170
Marsh 150
Penduline 152
Willow 151
Treecreeper 181
Short-toed 182
Tringa erythropus 74
glareola 76
nebularia 75
ochropus 74
stagnatilis 75
totanus 76
Troglodytes troglodytes 182
Turdus iliacus 186
merula 187
philomelos 186
pilaris 187
torquatus 188
viscivorus 185
Turnstone 66
Tyto alba 118

Upupa epops 123

Vanellus vanellus 61
Vulture, Bearded 105
Egyptian 105
Griffon 106

Wagtail, Ashy-headed 205
Black-headed 205
Grey 203
White 204
Yellow 205
Wallcreeper 180
Warbler, Barred 172
Cetti's 171
Dartford 178
Eastern Subalpine 174
Garden 172
Grasshopper 163
Great Reed 162
Icterine 158
Marmora's 176
Marsh 161
Melodious 159
Moltoni's 175
Moustached 160
Reed 161
Sardinian 174
Savi's 163
Sedge 160
Spectacled 177
Western Bonelli's 169
Western Orphean 173
Willow 169
Wood 167
Yellow-browed 167
Wheatear, Eastern Black-eared 197
Isabelline 196
Northern 196
Whimbrel 64
Whinchat 195
Whitethroat, Common 176
Lesser 173
Wigeon 25
Woodcock 71
Woodlark 154
Woodpecker, Black 131
Great Spotted 129
Green 132
Grey-headed 130
Lesser Spotted 130
Middle Spotted 128
Three-toed 127
White-backed 128
Woodpigeon 45
Wren 182
Wryneck 127

Yellowhammer 218

Zapornia parva 55